the joy of keeping
chickens

SECOND EDITION

the joy of keeping

chickens

THE ULTIMATE GUIDE TO RAISING POULTRY FOR FUN OR PROFIT

Jennifer Megyesi

photography by Geoff Hansen

Skyhorse Publishing

Copyright © 2009, 2015 by Jennifer Megyesi
Photography copyright © 2009, 2015 by Geoff Hansen

Skyhorse Publishing books may be purchased in bulk at special discounts
for sales promotion, corporate gifts, fund-raising, or educational purposes.
Special editions can also be created to specifications. For details, contact the
Special Sales Department, Skyhorse Publishing, 307 West 36th Street,
11th Floor, New York, NY 10018 or info@skyhorsepublishing.com.

Skyhorse® and Skyhorse Publishing® are registered trademarks of Skyhorse
Publishing, Inc.®, a Delaware corporation.

Visit our website at www.skyhorsepublishing.com.

10 9 8 7 6 5 4 3 2 1

Library of Congress Cataloging-in-Publication Data
Megyesi, Jennifer Lynn, 1963-
The joy of keeping chickens : the ultimate guide to raising poultry for fun or
profit / Jennifer Megyesi ; photography by Geoff Hansen.
p. cm.
Includes bibliographical references and index.
1. Chickens. 2. Poultry farms. I. Title.
SF487.M42 2009
636.5—dc22
2008047797

Cover design by LeAnna Weller Smith

Print ISBN: 978-1-63220-467-7
Ebook ISBN: 978-1-63450-026-5

Printed in China

For **Bradford**, who at eight months, pointed up to the dark sky outside the farmhouse and muttered in baby breath, "Moon," and who I know will reach for and capture the stars.

Peggy, a Polish Crested hen, basks in early sunlight at Fat Rooster Farm in Royalton, Vermont.

CONTENTS

"I like chickens because:

#1.   I like eating eggs.

#2.   I like holding chicks.

#3.   I like feeding chickens.

#4.   I like collecting chicken eggs.

#5.   I like washing chicken eggs.**"**

—Bradford Jones, age nine

The Joy of Keeping Chickens

FOR TWO WEEKS NOW, I have been unable to come up with a format to begin my talk about the joy of keeping chickens. It's a topic that I have been invited to present at the children's conference at the annual Northeastern Organic Farming Association meeting in Randolph, Vermont (NOFA-VT). This year, the conference, with its ever-growing popularity, has stretched into a two-day affair, and I have been asked to provide a one-hour presentation about chickens for children.

This seems like a no-brainer, for someone who has raised chickens almost her whole life, for someone who is commonly, and kindly, referred to as the crazy chicken lady.

But now, I have angst. What will this audience gain from my talk? They range from six to twelve years in age, and some may already know a lot about raising chickens.

I make up a list of questions, a quiz for them to take, about who lays eggs, and who crows good morning. Then I remember that they are children, and filling out forms and taking tests are chores done best at school, not in fun forums at an educational conference.

So I pack up Henry, the 10-pound Barred Plymouth Rock rooster, who boards with us annually because his Vermont summer parents have no safe winter haven for him to stay in while they travel to warmer climes. Also packed are Danny and Margaret Suzanne (Peggy Sue, for short), the Polish Cresteds, who can barely see through their headdresses of scattered, wild feathers that make them look like voodoo dolls. Then there are the pair of Old English Game Bantams, proud and glittering with their emerald-black feathers. There are my two prized White Silkies, whose sole existence on the farm is just for sheer grin in the morning at seeing their fur-like plumage while I rush around doing the rest of the barn chores. Silkies don't lay many eggs, and their black-skinned bodies put off the best of cooks unless they specialize in cooking broiled Chinese black fowl. Last, but not least, I add a Russian Orhloff, not yet recognized by our own American Poultry Association, but beautiful and worth showing off anyway.

The cages are covered with blankets on this cold, bleak February day; these birds have spent a pampered winter in a tight barn that never goes below freezing and stays as humid as a day spa. They are packed like sardines into crates in the back of the truck, but by the time we reach the conference, the Silkies have found their freedom, bouncing around in the back of the truck, looking for a way out of fame.

As I unload the crates, I begin to formulate the final parts of my presentation. The blankets will stay over the cages to draw the children to the talk. At the very last moment, I will pluck Henry and the two docile Silkies from their cages and plop them onto the tarp that has been spread to catch their inevitable offerings.

Henry's presence draws even the adults. Children, some of these weighing no more than three times what he does, take turns picking him up and cradling him like a doll. The Silkies are cooed at. The Old English are rocked into submission and fed vast amounts of grain so that everyone is astonished at the burgeoning size of their stomachs. Danny and Peggy Sue are gawked at, each becoming a little self-conscious in the end, so I eventually retire them to their cages.

Then a wonderful thing happens. The children forget that I am there, and they pair off with their respective chickens. They swap stories of chickens that they have had, or they wonder together at the feel of a Silkie's feathers and the warmth of Henry's toothed comb. They pet the chickens, talk to them, and offer them water and food. There are chicken kisses and hugs and chicken–human dialog. And while I talk about how chickens "sweat," and who lays eggs, and who does the barnyard crowing, it's the joy of keeping chickens taking over, and really I haven't needed to stress out about a thing.

RIGHT Laura Vaillancourt, 8, of Bethel, Vermont, holds a White Silkie rooster the author brought for a workshop at the NOFA-VT Annual Conference in Randolph Center, Vermont.

Climbing to the top of the gutter cleaner's conveyor belt, a rooster greets the Eastern sky before attempting to fly out into it at Fat Rooster Farm in Royalton, Vermont. Instead of soaring over the field below, the chicken went into a squawking free fall.

> **"A man who has one foot in industry and the other on the soil is about as secure as he can be in this world."**
> **–Henry Ford**

1

Why Keep Chickens

My husband and I are at a rest stop on I-95 in Maine, eating our lunch. We work as biologists for the Department of the Interior, and lately, I have begun to realize how far from the land and agriculture I have become, nestled in my ivory tower of science and information. I am facing away from the parking lot, toward the woods, when Kyle says, "Oh, no, don't look." And, of course, I have to. I turn to see a tractor trailer stacked with crates of chickens, the tarps on either side of the truck rolled up to reveal the battery cages, stuffed with white chickens with large, floppy combs. At first I think they must be roosters headed for slaughter, and of course, I insist that we must investigate. At the truck, I discover that it is loaded with cages of laying hens (probably just over one year of age), destined to be soup. The cages are stacked one upon another, at least 8 feet high, and in each, several chickens huddle. I think I can count at least 15 birds in every cage. Their beaks have been cut so they can't cannibalize each other in their tiny cages or eat their own eggs, and they have toenails that have curled upward for lack of any chance to scratch in the earth and wear them down. They are ragged and spent, and their mouths are open, gasping for air in the rare Maine heat. On the other side of the truck, it is much worse. On this side the flaps have been open to the sun, and there are chickens in every cage that have succumbed to this unaccustomed heat, lying dead in the cages with their compatriots.

The horror of this suddenly hits me. These laying hens are done with what they have been contracted to do and are headed for the soup factory on a crowded truck. Why haven't I ever considered this when I buy my eggs? Certainly, I am concerned with veal calves, kept in their climate-controlled stalls, filled with antibiotics and grain for veal production, their big, brown eyes blinking, and that's why I won't eat commercially raised veal. But how did I think that most eggs appeared in those cartons, anyway?

I'm covering my face with my hands now, and my husband is trying to quietly usher me back to the car, when a big, tall man in a pinstriped shirt that reads "DeCoster Egg Farm, FLOYD" approaches me. I ask him where the chickens are headed, and he tells me that they are going to be made into soup or broth. Then, between the stacks of cages, I see a tiny white body. She has found a way out, and has mashed herself in between the rows, concealed beneath the truck's flaps. "You can have her if you like," he says to me, possibly concerned that he has met someone from PETA. My husband crosses to the other side of the truck, and I stay hidden, until the little hen dashes from her cave. I snatch her up and whisk her away. She is terrified and shudders, cackling alarm. Her name becomes Floyd, and she is my pledge to stay aware of why I have decided that I must raise chickens. If I am to eat meat and eggs, then I need to take responsibility for caring for them in a humane manner, from start to finish.

It took two months to teach Floyd how to eat and drink on her own, having been used to mechanized delivery of nourishment. After her initial adjustment, she laid eggs every day for nearly a year. She took two months off to molt her spent feathers, then resumed production. She lived almost three years at Fat Rooster Farm, when she finally died of cancer, a common demise of the commercial strains, developed for production, not longevity.

Floyd shocked me back into reality. The convenience of purchasing eggs at the local shopping market had completely removed me from the fact that something alive had laid those eggs. I had not been concerned *with how the eggs were produced, how the hens were cared for, and with the final disposition of the creatures that had laid the eggs.*

In the spring, my father would sometimes give my sisters and me a choice: either we three could go to church with Mom, or we could pile into the station wagon and head over to Teeny's Tiny Poultry Shop, a half hour's drive from our house. Both my younger sister and I would be out of our Sunday finest and in the car before the chill of the morning air had left our rural Vermont valley. My middle sister, the pragmatist, would weigh her decision carefully, but in the end, the entire family made the trip.

The adage "April chicks bring September eggs" works well in our northern clime, and the chicken run to Teeny's was a sure sign

that spring's arrival was close, and green grass and dandelions were not far off.

Before my sisters and I were born, my parents lived in California. One night at a party in San Francisco, they saw a calendar of Vermont on the wall. Eventually, the memory of those scarlet maples and covered bridges guided them east. In our first home in Vermont, I slept in a dresser drawer, the only crib my parents could afford, and it worked just fine. Dad eventually became a teacher at the high school, and Mom raised us kids and was a registered nurse at Middlebury College. Both of

LEFT: In a cardboard box in the living room, a newly hatched chick sits under a heat lamp after emerging from an egg set in an incubator.

While the meat birds follow with great interest, the author walks with a bucket of grain out to a portable coop.

them imparted a deep sense of wanting to be involved with the living world and to know how to nurture and nourish ourselves, life skills that many of us have now forgotten or don't know how to teach our children. Having an iPod or an iPhone with the newest, latest, greatest technology is more important than learning where the food we eat comes from, whether chickens can taste their food, or whether a Leghorn hen is used more often for egg production than for meat production.

Chicken Vital Signs

- *Body temperature: 105 to 107 degrees Fahrenheit*
- *Respiration: Rooster, 18 to 20 breaths per minute; hen, 30 to 35 breaths per minute*
- *Heart rate: 280 to 315 beats per minute*
- *Average age before egg-laying commences: 6 months*

Our food is being outsourced, in front of our eyes, at the expense of our health and ability to reliably ensure its safety and quality. Personal freedom to choose products other than from large corporations like Tyson, Pilgrim's Pride, and Gold Kist, who control 48% of the chicken meat production in the United States and account for 85 million broiler sales in the United States weekly, is slowly being taken away. In fact, in 2006, fully 15.7% of organic grocery sales in the United States were made at Wal-Mart. Our ability to fend for ourselves and to feed ourselves is being replaced with the idea that progress and civilized culture can only be achieved outside of the home, away from the chicken yard, far from the kitchen, closer to the frozen selection of ready-to-eat meals.

And although the outright joy of collecting the family flock's warm, fresh eggs to make an omelet far outweighs how happy it makes me feel to peel back the wrapper on a frozen, microwaveable breakfast burrito, the psychological considerations of raising food oneself go even deeper. Keeping chickens is as rewarding as growing sun-ripened tomatoes; they are the livestock most likely to be kept by the back-to-the-landers, the homesteaders, and the dreamers of a place in the country. They have been domesticated for thousands of years, and even the Egyptian pharaohs used artificial incubation to rear their own chicks. Chickens are relatively hearty, require little living space, can produce both meat and eggs, and are inexpensive to purchase.

Americans annually consume 87 ½ pounds of chicken meat and 256 eggs per capita, a number that has continued to grow leaps and bounds above the average consumption of beef, pork, or lamb. Both chicken and eggs are important sources of high-quality protein, vitamins, and minerals. Chicken meat is higher in protein than red meat, is second in vitamin D content only to fish oil, and is much easier to raise and process than the latter two.

Darwin believed that *Gallus domesticus*, the modern-day chicken, descended from the Red Jungle Fowl (*Gallus gallus*) in Southeast Asia some 3,000 years ago. Recent DNA evidence supports this theory, though some dispute that chickens such as the Cochin and Brahma may have descended from other breeds near China, Vietnam, or India. Domestication occurred with the advent of cock fighting within the Roman and Greek armies. After conquering territories and moving on, surplus birds left behind continued to multiply, and inbreeding and isolation led to many variations and different breeds. In the mid-nineteenth century, an interest in fancy poultry was at its height in England. Queen Victoria kept a royal poultry yard and was given a pair of Cochins (formerly called Shanghais) as a gift. Poultry fever hit when the birds were displayed at the Show of the Royal Dublin Agricultural Society in 1846, and everyone present wanted to purchase them. When cockfighting was banned there in 1849, poultry exhibitions and clubs sprang up to take the place of the fighting ring, so poultry shows increased in popularity.

The order of arrival of chickens to the New World has been much disputed. Columbus brought chickens on his second voyage in 1493, though scientists have discovered ancient remains of chickens in Chile that could date back as early as 1300 AD. This would point to the Polynesians as introducing them first, perhaps as a source of food. Early American colonists imported chickens to Jamestown, Virginia, in 1607 to use primarily for eggs and cockfighting. (Americans did not consider chickens a good source of meat until sometime later.) The poultry craze hit the United States in the mid-1800s after the

CHICKEN BASICS

All chicken flock owners should familiarize themselves with basic chicken anatomy. It is useful when comparing other specimens of the same breed, and it is crucial to know what a normal chicken looks like in order to catch potential health problems before they are spread to the rest of your flock. Chapter Ten covers specific diseases and conditions that can easily be detected in the home flock given a simple understanding of a chicken's normal anatomy and physiology.

- The comb should be full and blemish-free in mature birds, indicating no frostbite, no underlying disease, and no scars from being overly aggressive toward other chickens.
- The eyes should be bright, clear, and alert.
- Nostrils should be free of discharge or crusty material. There should be no sounds emitted during normal respiration.
- The wattles should be rounded and undamaged—damage being another indication of fighting or exposure to extreme elements or disease. Check around the base of the wattles for parasites and their eggs. Lice will lay eggs in clusters that look like brittle clumps of whitish-colored debris.
- The plumage should be in good condition–bright, lustrous, sleek, and well-maintained.
- Check the vent for parasites as well; manure-caked feathers here may indicate disease or advanced age in birds.
- The breastbone should be straight, free of abscess, and well-feathered.
- The legs should be well-scaled and smooth. Inspect the toenails for debris. One indication of advanced age is the spur in the rooster; these are not well-formed until the second year of life.

① ←——

⑬ ←——

⑫ ←——

① Beak
② Head
③ Back
④ Upper saddle
⑤ Tail
⑥ Lower saddle
⑦ Vent
⑧ Wing
⑨ Fluff
⑩ Toes or Claws
⑪ Shank
⑫ Keel or Breast
⑬ Hackle
⑭ Chick

Chicken Scratch

first poultry exhibition held in Boston in 1849 attracted more than 10,000 people to the show. The American Poultry Association was formed in 1873, and one year later, the *American Standard of Excellence* was adopted. After World War II, poultry business became industrialized, and improvements were made by crossbreeding to produce superior egg layers and meat birds. With the rise in commercial value for chickens, many breeds, now referred to as heritage breeds, were eliminated from flocks if they weren't able to compete with the crossbreeds for eggs or meat production. Of the 55 chicken breeds listed by the American Livestock Breeds Conservancy (ALBC), an organization concerned with preserving livestock diversity, more than 26 breeds are listed as critical or threatened in the United States Besides the loss of diversity to guard against catastrophic illness in our domestic poultry population, keeping heritage breeds to preserve their unique characteristics alone may be motivation enough for you to keep chickens.

By keeping chickens on your farm or homestead you'll achieve a safe, high-quality food source for the family. There are other joys that they can bring, like the pleasure my son shows when he has discovered a nest wriggling with newly hatched chicks, or when he quietly watches the birds scratch and pick at the soil for food, marveling at their many shapes and colors, and laughing at their quirky chicken antics

For me, chickens are a link to childhood. They're responsible for filling me with a wonder of living things, a means toward self-sufficiency and a lesson in compassion for living creatures that holds much more weight than just having someone tell me that we're entitled to dominion over all other living things because we're humans.

Chickens are curious creatures, easily sloughed off as "bird-brained" and inconsequential in the grand scheme of living things; certainly they're not as smart as dogs or ponies. And yet, like watching aquarium fishes swim in a tank, watching chickens in the yard and being responsible for their care and comfort has taught me a respect for caring and preserving life, as I know it.

Whether you decide to keep chickens for eggs and meat, for show, or just for companionship, learning to care for them is easy and rewarding, providing that you know some basics. This book should give you what you need to get started building and keeping a flock; choosing breeds that will fit your goals; starting and raising chicks; housing and protecting your flock from predators; hatching and incubating eggs; and gathering, storing, using, and marketing the eggs and meat that you produce.

RIGHT: Going out for fresh air, a White Leghorn hen steps outside of the coop. The farm's chickens are free to roam the barnyard in the summertime.

Eggs gathered from the nesting boxes are
set aside in a basket before they are washed
for customers.

" There are as many ways of looking after poultry as there are fashions in childrearing. Pick a system that suits you and enjoy yourself, ignore the avian mother-in-laws tut tutting away. "

–Francine Raymond, *The Big Book of Garden Hens, 2001*

2

Starting Your Poultry Enterprise

Early on, when my husband and I first began to raise meat birds, I foolishly tried to spare some of the friendliest ones and keep them as pets. We would fondly refer to these birds as the "Jabbas," after the grotesque character in Star Wars—their sole purpose in life seemed to be eating, sleeping, and looking slovenly. I knew better than to become attached to what I was raising for meat, but I couldn't help noticing their individual personalities and their docility. I named one Forrest Gump and another Wendell Berry, after personal heroes. Both chickens were horrendous choices for companions, gullible, devoted, even cuddly, but both were gi-normous, fat beasts that could hardly move.

TOP: While a Tunis cross ewe watches, the author checks on newly delivered chicks she had just placed in a brooding box. A heat lamp keeps the chicks warm as they adjust to their new surroundings.

ABOVE: Once their son grew out of his playhouse and swing set, Carol Steingress and Rick Schluntz converted it into a coop, left, and a winter run for their half-dozen laying hens in Windsor, Vermont. Before they got chickens, Steingress checked to see if the town's zoning laws prohibited raising chickens in town. It did not, but she and her husband don't keep a rooster to keep the peace in the neighborhood.

Forrest died of a heart attack after spending nearly eight months as a teacher liaison for migrant workers' children, where he tried hard to convince the kids that farm animals were friendly. Wendell held on a little longer, but eventually succumbed to bone cancer at the ripe age of two.

When I asked my husband if we could posthumously name our newly acquired farm White Rooster Farm, he looked at me with alarm. "I can't possibly think why we should pay tribute to these genetic mutants, can you?" So instead, we agreed upon Fat Rooster Farm—a place where we could endeavor to raise our chicken meat and eggs responsibly, with the full understanding of these creatures' lives and their worth.

We decided to raise the birds using organic methods, to look for varieties that thrived on pasture and wouldn't self-destruct at a young age. We knew that it would be a challenge. Our laying hens would not lay just brown eggs, but white, pink, green, and mahogany-colored ones. The meat birds would grow more slowly and have bodies that resembled chickens of the 1950s in many respects. Their breasts would have keels (the long, sharp bone where their breast muscles attach under their bellies) and their legs would be long and sturdy to support them in their search to find foods rich in vitamins, minerals, and fatty acids.

It wouldn't be as rewarding right up front to raise meat and eggs this way, but we would be able to say truthfully that we'd tried to raise the birds humanely and kindly. And eventually, we gained the respect and following of other fellow meat and egg eaters, who value slow food and the knowledge that the animals have been treated compassionately.

Some Considerations Before You Get Started

Local and Federal Regulations

Local laws, such as zoning and other ordinances, may end your chicken enterprise before it takes flight. Some farming magazines have been advertising "Stealth Coops," mini-pens to conceal your poultry in more urban settings. Besides looking uncomfortable for the chickens (there doesn't seem to be much space for more than two or three chickens, nor are there sufficient locations for nesting, roosting, and foraging), the idea of concealing your chickens from your neighbors or from town and city officials seems like it will only lead to trouble.

The best way to avoid conflicts is to check with the county zoning board, your extension agent, or the state agricultural department. And after all the rules check out, consult your neighbors with your plan. Are they okay with the accompanying baggage associated with keeping chickens—noise, or

" The fact that zoning in towns allows residents to raise a barking, crapping dog the size of a small elephant, but not four hens for a steady fresh egg supply shows just how lacking in common sense we have become as a society. The indictment against town chickens stemmed originally from roosters crowing at dawn and waking neighbors. The easy solution to that problem is to raise only hens or butcher the roosters before they mature enough to crow all the time. "

—GENE LOGSDON in *Chicken Tractor*, by Andy Lee, 1994

ABOVE: Cloë Milek pets a Dark Brahma Cochin chick in her 2/3-acre in-town yard in Windsor, Vermont. Milek grew up on a farm with chickens and has been raising them with her fiancé for the past seven years. "It's like therapy," she said of raising chickens. "It's very relaxing."

ABOVE: Two Barred Plymouth Rock pullets, left, and a Partridge Plymouth Rock pullet roam the yard of Carol Steingress and Rick Schluntz. The couple have been raising hens for eggs in their small yard for the past 10 years.

Chicken Scratch

Do I need a rooster to have my hens lay eggs?
No. The rooster is needed if you want a fertile egg but isn't necessary for a hen to lay eggs.

Do hens lay more than one egg a day?
No. In fact, they will average less than an egg a day in many cases, and their productivity depends on many factors, including age, diet, weather conditions, daylight hours, and breed.

Are all roosters mean?
No, but some breeds have a proclivity toward mean roosters (see breed chart on page 38).

Do roosters only crow at sunrise?
No—they will crow when startled, when they are establishing hierarchy, or simply to announce their territorial presence.

an enticement for the dog to chase? Perhaps ruffled neighborhood feathers can be smoothed with the promise of fresh eggs and meat, or the reward of manure high in nitrogen for their garden.

What Type of Chicken Project Interests You?

Before acquiring birds, figure out what you want to accomplish. Chicken fever (a term used loosely to describe someone acting like they're in a candy store, overwhelmed with the choices of sweets available and ending up with a bellyache instead of a wise choice of one type of candy) is hard to avoid when so many interesting and colorful breeds exist. In reality, each breed of chicken can differ in its specific requirements for food, water, and shelter. A mixed flock may not necessarily do well together. The type of husbandry you have in mind will determine what breed of chicken you decide upon. Answering these questions beforehand will help you determine how chickens will become part of your plan.

- Will the birds be used as a source of eggs and meat for the homestead?
- Do you want to develop a small-scale commercial enterprise?
- Is it easier to market fresh eggs or meat near your farm?
- Are there poultry processing facilities near to you, or do you plan to butcher them yourself?
- Will they be confined to a poultry house, or do you wish to raise them on pasture?
- Are you planning to raise purebred, show-quality poultry?

Generally speaking, then, there are four broad enterprises for raising chickens: for egg production, for chick production to sell locally or to hatcheries, for meat production, or as a hobby (for show or companionship).

How Chickens Are Described

As with other animals and plants, specific terms are used to describe chickens. Breed, class, variety, and strain identify each of the

ABOVE: Black Star cross hens perch on a bale of hay in the greenhouse at Luna Bleu Farm in South Royalton, Vermont. Farmers Tim Sanford and Suzanne Long house the chickens in unused greenhouses in winter to give the chickens plenty of light and warmth.

different chickens. For a complete list of breed attributes, including temperament, egg color, hardiness and their primary uses, refer to the Chicken Breed Chart and Key (page 38).

BREED

Breeds of chickens are grouped together according to their size, shape, plumage, number of toes, color of their skin, etc. When mated together, individuals within a breed will produce chicks that share the same characteristics as their parents. Breed recognition can be different, depending on which poultry organization you consult. For example, the Marans is a French breed that is not recognized by the APA (American Poultry Association) because of its inconsistency in type (some Marans individuals in the United States have feathered legs, while others do not), but it is recognized as a distinct breed by the Poultry Club of Britain (PC). On the website www.feathersite. com, there are hundreds of breeds listed that occur worldwide. The APA only recognizes 53 large fowl and 61 breeds of bantams.

CLASS

Breeds are subdivided into classes. In large breeds, classes indicate their origin: American, Asiatic, English, Mediterranean, Continental, and Other (which includes Oriental). Bantam breeds are classified according to characteristics, like comb shape, or presence of feathering on the legs.

VARIETY

Varieties describe breeds based usually on plumage color, but also on comb style or feathering. For example, the Leghorn, in the Mediterranean class, has 12 varieties of large breeds.

STRAIN

A strain is a term that refers to a line of birds that have been bred for specific characteristics. In the show arena, strains are developed from a single breed for characteristics that are thought of as typical or "typy" and are considered superior by the owner and by fanciers of this same breed. There can be several different strains within the same breed. Commercial strains are often hybrids, sometimes having parents of different breeds. These strains are developed for superior production of either eggs or meat. A Cornish-Rock cross is a popular meat hybrid bred to grow heavy breast and thigh meat in a short amount of time. The Black Sex-Link is a laying hen that is crossed

> " However reluctant those concerned with poultry may be to acknowledge the fact, it is not the less true that most old women who live in cottages know better how to rear chickens than any other persons; they are more successful, and it may be traced to the fact that they keep but few fowls, that these fowls are allowed to run freely in the house, to roll in the ashes, to approach the fire, and to pick up any crumbs or eat all the morsels they find on the ground, and are nursed with the greatest care and indulgence. "
>
> —SIMON SAUNDERS, from *Domestic Poultry: Being a Practical Treatise on the Preferable Breeds of Farm-Yard Poultry*, 1868

with a Barred Plymouth Rock and a Rhode Island Red that can be distinguished as a female from its plumage as a chick.

Foundation versus Composite Breeds

Poultry literature will often refer to chickens as being either foundation or composite breeds. A composite poultry breed is somewhat akin to a crossbred dog such as the Labradoodle, created using a purebred Labrador Retriever (a dog known for its easygoing personality) and the Standard Poodle (a dog known for its brains).

A foundation breed is a very old breed of chicken with distinct characteristics, such as the Dorking, with its five toes. This breed, along with Houdans and Asiatic breeds, was used to develop the composite breed in northern France called the Faverolle. The Faverolle was well adapted to battery cage production, where laying hens were confined in small cages for maximum egg production, and was a good source of winter eggs in the late 1800s.

CHICKEN PERSONALITIES

Behavioral traits among chickens vary widely. The American class contains breeds such as the Rhode Island Red, the Plymouth Rock, and the Jersey Giant that are generally docile, are cold tolerant, and can produce eggs and meat on a small scale. Most of the breeds in the Mediterranean class, like the Leghorn and Minorca, are flighty, smaller bodied, less cold tolerant, but more efficient in converting feed to egg production. The Dorking and Java are both excellent meat breeds, capable of foraging for themselves and requiring less grain, but they will mature more slowly than the Cornish Rock hybrids that are almost entirely dependent on being fed high-quality mixed rations. Dual-purpose breeds refer to those birds that produce enough good-quality meat and sufficient numbers of eggs that they could fit into small-scale operations for home use. It will be important to evaluate the traits of each breed to determine which would be most suitable for your enterprise. Several hatcheries offer assortments of chicks, so you can experiment in your first year or two. It might also be useful to visit neighboring flocks to discuss the pros and cons of the breeds with the flock owner.

1. A Buff Orpington pullet roosts in the enclosed pen outside the coop owned by Cloë Milek and Karl Hanson on their 2/3-acre lot. The couple has been raising 20 hens for eggs for the past seven years. They have avoided keeping a crowing rooster to keep the peace with their neighbors.

2. A Black Star cross hen basks in the late winter light at at Luna Bleu Farm. The farm raises 100 hens for laying eggs and another 60 pullets are growing for their turn in the nesting box.

3. A Rhode Island Red hen, left, and a Golden Comet hen look at each other outside of the coop at Back Beyond Farm. Farmer Ray Williams prefers the Comets to the Rhode Island Reds, whom he has seen breaking eggs in the coop.

4. While Pistol the Old English Game bantam gives herself a dust bath, one of her chicks explores outside the coop belonging to Cloë Milek and Karl Hanson. Pistol hatched the chicks from fertile eggs Hanson got from a nearby farm.

3

4

5

6

5. Rosie, a three-year-old Black Sex-Link hen, roams the backyard pen at the home of Geoff Hansen and Nicola Smith. When Hansen bought three pullets from a local farmer for their new coop, Rosie was included as a mentor to teach them the egg-laying ropes.

6. Danny, a Polish Crested rooster, is one of a menagerie of birds—including chickens—kept by the author.

7. A Cornish Rock cross meat bird rests at Back Beyond Farm in Tunbridge, Vermont. Farmer Ray Williams said customers prefer the variety's abundance of breast meat; the chickens weigh about seven pounds when packaged for sale.

8. With a puff of feathers around its neck and the dark eggs it lays, a Russian Orloff hen is a unique bird. The author was given the bird by someone who was moving away.

Chicken Breed Chart

Breed	Class	Type	Recognized by	Varieties	Productivity	Egg Color	Meat	Sport	Ornamental	Temperam
Ameraucana	Other (Miscellaneous)	l, b	APA	8	3	other	1		*	d
Ancona	Mediterranean	l, b	APA, PC	1	3	white	1		*	f
Andulusian	Mediterranean	l, b	APA, PC	1	2	white	1		*	f
Appenzeiler or Appenzell	Developed in Switzerland	l	PC	2	2	white	1			f
Araucana	Other (Miscellaneous)	l, b	APA, PC	5	2	other	1		*	d
Aseel or Asil	Other (Oriental)	l	APA	6	1	brown	2		*	d
Australorp	English	l, b	APA, PC	1	3	brown	3			d
Barnevelder	Continental	l	APA, PC	1 (APA)	3	dk brown	2		*	d
Brahma	Asiatic	l, b	APA, PC	3	3	brown	3		*	d
Buckeye	American	l, b	APA	1	2	brown	2			d
Buttercup	Mediterranean	l, b	APA, PC	1	3	white	1		*	f
Campine	Continental	l, b	APA, PC	2	3	white	1		*	f
Catalana	Mediterranean	l, b	APA,	1	3	white	3			f
Chantecler	American	l, b	APA	2	3	brown	3			f
Cochin	Asiatic	l, b	APA, PC	9	2	white	2		*	d
Cornish	English	l, b	APA, PC	5	1	brown	3			d/a
Crevecoeur	Continental	l, b	APA, PC	1	3	white	1		*	f
Cubalaya	Other (Oriental)	l, b	APA	3	1	white	2	*	*	a
Delaware	American	l, b	APA	1	3	brown	3			d
Dominique	American	l, b	APA, PC	1	3	brown	3			f
Dorking	English	l, b	APA, PC	6	3	white	2			d
Faverolles	Continental	l, b	APA, PC	2	3	other	3		*	d
Fayoumi		l	PC	2	3	white	1			f
Hamburg	Continental	l, b	APA. PC	6	3	white	1		*	f
Holland	American	l, b	APA	2	3	white	1			d
Houdan	Continental	l	APA, PC	2	2	white	2		*	d
Java	American	l, b	APA, PC	2	2	brown	3			d
Jersey Giant	American	l, b	APA, PC	3	2	brown	3		*	d
La Fleche	Continental	l, b	APA, PC	1	3	white	1		*	f
Lakenvelder	Continental	l, b	APA, PC	1	3	white	1		*	f
Langshan	Asiatic	l, b	APA, PC	3	3	brown	2		*	d
Leghorn	Mediterranean	l, b	APA, PC	18	3	white	1		*	f
Malay	Other (Oriental)	l, b	APA, PC	6	1	brown	1		*	d
Marans		l, b	PC	3	3	dk brown	1			d
Minorca	Mediterranean	l, b	APA, PC	7	3	white	1		*	f
Modern Game	Other (Game)	l, b	APA, PC	9	1	white	1		*	a
Naked Neck	Other (Miscellaneous)	l, b	APA, PC	4	2	brown	1		*	d
New Hampshire	American	l, b	APA, PC	1	3	brown	1		*	d / a
Old English Game	Other (Game)	l, b	APA, PC	15	1	white	1	*	*	a
Orloff		l	PC	4	2	brown	1			d
Orpington	English	l, b	APA, PC	4	3	brown	3			d
Penedesenca		l		4	3	dk brown	3			d
Phoenix	Other (Oriental)	l, b	APA, PC	2	1	white	1		*	d
Plymouth Rock	American	l, b	APA, PC	7	3	brown	3		*	d
Polish	Continental	l, b	APA, PC	10	1 to 3	white	1		*	d
Redcap	English	l, b	APA, PC	1	3	w	1		*	f
Rhode Island Red	American	l, b	APA, PC	2	3	brown	3			d
Rhode Island White	American	l, b	APA, PC	1	3	brown	3			d
Shamo	Other (Oriental)	l, b	APA, PC	5	2	brown	3	*	*	d
Spanish	Mediterranean	l, b	APA, PC	1	3	white	1		*	f
Sultan	Other (Miscellaneous)	l, b	APA, PC	1	1	white	1		*	f
Sumatra	Other (Oriental)	l, b	APA, PC	1	1	white	1		*	a
Sussex	English	l, b	APA, PC	3	3	brown	3			d
Welsummer	Continental	l, b	APA, PC	1	3	dk brown	3			d
Wyandotte	American	l, b	APA, PC	9	3	brown	3		*	d
Yokohama	Other (Oriental)	l, b	APA, PC	2	1	brown	1		*	d/a

Hardiness	Broodiness	Weight: Cock/Hen	Comb Type	Skin Color	U.S. Availability	Status	Notes
c	1	6.5 lbs/5.5 lbs	pea	s	c	c	most "Easter Egg' chicks are of mixed breeding
c	3	6 lbs/4.5 lbs	single	y	u	r	
c	3	7 lbs/5.5 lbs	single	w	u	c	
c	2	4 lbs/3.3 lbs	v-shaped	w	u	r	
c	1	5 lbs/4 lbs	pea	y	r	s	purebred tufted birds are very rare
c	1	5 lbs/4/lbs	pea	w	u	c	
c	1	8.5 lbs/6.5 lbs	single	w	c	r	excellent dual-purpose breed
w	2	7 lbs/6 lbs	single	y	u	c	considered by some an excellent dual-purpose breed
c	1	12 lbs/9.5 lbs	pea	y	u	w	slow maturing, hardy in cold or heat
c	1	9 lbs/ 6.5 lbs	pea	y	r	c	
w	3	6.5 lbs/5 lbs	buttercup	y	r	c	extremely flighty
c	3	6 lbs/ 4 lbs	single	w	u	c	
w	3	8 lbs/6 lbs	single	y	r	c	good dual-purpose breed
c	2	8.5 lbs/ 6.5 /bs	cushion	y	r	c	extremely rare, even in Canada
c	1	11 lbs/ 8.5 lbs	single	y	c	w	very good mothers
c	2	10.5 lbs/ 8 lbs	pea	y	u	w	now used to cross with Plymouth Rocks for meat birds
t	3	8 lbs/ 6.5 lbs	v-shaped	w	r	c	
w	1	6 lbs/ 4 lbs	pea	w	u	w	just recently popular
c	2	8.5 lbs/ 6.5 lbs	single	y	r	c	good dual-purpose breed
c	1	7 lbs/ 5 lbs	rose	y	u	w	
c	1	9 lbs/ 7 lbs	single	w	r	c	extremely rare
c	2	8 lbs/ 6.5 lbs	single	w	u	c	gaining in popularity
c	3	4 lbs/ 3.5 lbs	single	w	u		not recognized by APA
c	3	5 lbs/ 4 lbs	rose	w	u	w	not tolerant of closed confinement; economical eater
c	1	8.5 lbs / 6.5 lbs	single	y	u	c	
c	2	8 lbs/ 6.5 lbs	v-shaped	w	u	c	
c	1	9.5 lbs/ 7.5 lbs	single	y	r	c	one of the oldest U.S. breeds
c	1	13 lbs/ 10 lbs	single	y	u	w	slow maturing, not an economical eater
c	3	8 lbs/ 6.5 lbs	v-shaped	w	r	c	very rare
c	3	5 lbs/ 4 lbs	single	w	r	r	early maturing
c	2	9.5 lbs/ 7.5 lbs	single	w	u	r	
c	3	6 lbs/ 4.5 lbs	single, rose	y	c		economical eater, very flighty
c	2	9 lbs/ 7 lbs	strawberry	y	r	c	
c	2	8 lbs/ 7 lbs	single	w	r		not recognized by APA, but becoming popular
c	3	9 lbs/ 7.5 lbs	single, rose	w	u	w	combs can be susceptible to frostbite
c	1	6 lbs/ 4.5 lbs	single	y	u	s	male's comb is dubbed for exhibition
c	3	8.5 lbs/ 6.5 lbs	single	y	u	u	
c	1	8 lbs/ 6.5 lbs	single	y	c	w	
c	1	5 lbs/4 lbs	single	w	c	s	male's comb is dubbed for exhibition
c	3	7 lbs/6 lbs	walnut	y	r		very rare, dropped from APA
c	1	10 lbs/8lbs	single	w	c	recov	only buffs are common
w	3	5 lbs/4lbs	carnation	w	r		very rare; just recently imported to U.S.
t	2	5.5 lbs/ 4 lbs	single	y	u	s	special housing, short-lived
c	2	9.5 lbs/7.5 lbs	single	y	c	recov	very popular
c	3	6 lbs/4.5 lbs	v-shaped	w	c	w	varies widely in egg production
c	3	7.5 lbs/6 lbs	rose	w	r	c	very rare
c	2	8.5 lbs/6.5 lbs	single/rose	y	c	recov	very popular, but purebreds are rare
c	2	8.5 lbs/6.5 lbs	rose	y	c	w	
c	2	11 lbs/7 lbs	pea	w	u	w	
w	3	8 lbs/6.5 lbs	single	b	r	c	
w	3	6 lbs/4 lbs	v-shaped	w	u	s	
c	2	5 lbs/ 4lbs	pea	y	u	c	
c	1	9 lbs/ 7 lbs	single	y	u	r	
c	2	7 lbs/6lbs	single	y	u		gaining in popularity
c	2	8.5 lbs/6.5 lbs	rose	y	c	recov	generally docile, but can be aggressive
t	3	4.5 lbs/3.5 lbs	walnut or pea	y	u	s	requires special housing due to plumage

Chicken Breed Chart Key

The breed chart is a description of the breeds recognized by the *American Standard of Perfection* as well as some other popular breeds that are as yet unrecognized.

Class refers to the subdivision that large (standard) and bantam fowl fall into. Classifications for standard fowl by the American Poultry Association (APA) indicate their place of origin, while bantams are classified according to specific characteristics, such as comb style, whether they are sport breeds, and may be represented in several different classes depending on variety. The Poultry Club of Britain (PC) distinguishes breeds based on feather type (hard and soft) and body type (light and heavy). Unrecognized breeds are those for which standards of perfection have not yet been determined, but are still considered by some fanciers as distinct breeds.

Breed refers to a group of birds that have similar characteristics, such as size and shape.

Type refers to whether the breed is represented by both large (standard) and bantam body types: l = standard large fowl; b = bantam.

Recognized By indicates if the breed is recognized by the APA, the PC, or is unrecognized to date.

Varieties categories are based on feather color, feather placement, or style of comb. For example, the Plymouth Rock has seven varieties, based on plumage color.

Productivity refers to the egg-laying ability of the breed, relative to other non-commercial breeds: 1 = poor to fair; 2 = good; 3 = excellent. Highly subjective, productivity is easily affected by climate, age, diet, living conditions, and strain.

Egg Color ranges from white (light cream, chalk, to pure white) to brown (light brown to brown); and from dark brown (chocolate brown to mahogany brown) to other (pink, blue-gray, blue, green, or green-gray).

Meat indicates the breed's relative ability to produce a quality carcass within 8 to 20 weeks of age in comparison to other non-commercial breeds: 1 = poor to fair; 2 = good; 3 = excellent. Color of the skin should also be considered.

Ornamental indicates that certain strains of the breed have been developed for show rather than meat or egg production.

Sport indicates that the strains of the breed have developed for sport historically.

Temperament is classed as f = flighty, based on behavior or (as in some crested breeds) the ability to be startled because of plumage covering their eyes; a = aggressive, meaning protective of flock, behavior within the flock, or to humans; d = docile. All are relative classifications and may not address individuals within each breed.

Hardiness is classed as c = relatively tolerant of cold; w = tolerant of warmer climates; t = do best in temperate zones.

Broodiness rates the tendency to incubate laid eggs and set on them to hatching: 1 = good brooder; 2 = moderately broody; 3 = non-setter.

Comb Type is based on APA standards: single, rose, spiked rose, cushion, pea, walnut, strawberry, buttercup, or vee-combed.

Skin Color refers to the skin color underneath the bird's feathers: s = slate; b = bluish; w = white; y = yellow; b = black.

Availability in United States is based on twenty hatcheries in the U.S. (see appendix for list of hatcheries): c = common; u = uncommon; r = rare; na = not available readily.

Conservation Status refers to the breed's overall status, based on the American Livestock Breed Conservancy (ALBC): c = critical, with fewer than 500 breeding birds, and five or fewer primary breeding flocks in North America; r = rare, with fewer than 1,000 breeding birds, and seven or fewer primary breeding flocks in North America; w = watch, with fewer than 5,000 breeding birds and 10 or fewer primary breeding flocks in North America; s = study, breeds that are of interest but lack either definition or historical documentation; re = recovering, breeds that have recovered from the other categories but still need monitoring; co = common, breeds that are not considered by ALBC to be in danger of extinction.

How to Get Your Flock Started

Chicks can be hatched under a hen by saving eggs from your fertile flock (meaning there are roosters running freely with the hens). Fertile eggs can also be purchased and placed under a broody hen or in an incubator. Most commonly, chicks are purchased as day-olds that are mailed and delivered within 24 to 48 hours, but hatcheries usually require a minimum order to keep the chicks warm during shipping. Females that are close to beginning egg production, called started pullets, are probably most economical if you are interested in production as early as possible. In rural areas, local newspapers and agricultural magazines often advertise laying hens at a free or reduced cost. Show-quality birds are offered by breeders in fanciers' magazines, at poultry shows, or on the Web. Appendix Two lists specific poultry breed clubs that you can contact for more breed specific information; Appendix Four lists hatcheries that sell a variety of breeds. The *American Standard of Perfection*, published by APA (www.amerpoultryassn.com), should be consulted for specific characteristics for each breed recognized by the organization.

Before introducing new stock to an existing flock, the birds should be quarantined and given a clean bill of health from a veterinarian or a knowledgeable poultry producer. The National Poultry Improvement Plan (NPIP) maintains a directory of hatcheries and breeders that are enrolled in blood-testing programs for detection of several diseases that are contagious to poultry. The voluntary program was started in the 1930s as a means of eliminating pullorum disease from commercial poultry. Not all breeders are willing to wade through the red tape of bureaucracy, however, and you could always have your state extension agent do the testing on the birds that you are interested in purchasing (see Appendix Three).

Hatching Chicks with Broody Hens

One of the most exciting processes to observe is that of a hen transforming her eggs to fluffy chicks just by sitting on them for 21 days. Tracking the development of the embryo by *candling* the egg and watching the hen as she keeps vigil on the nest provides a magical quality to raising your own flock.

Before you can hatch the eggs, however, you need a hen that still has the maternal urge to become *broody*. A broody hen is one that has laid a nest or *clutch* of eggs and has begun sitting on them (setting) with the intention of hatching. Once the hen becomes broody, she will cease to lay eggs and will not resume laying until her chicks have reached independence (the age of independence varies according to breed). A broody hen will readily accept eggs that are given to her even if they have been laid by other hens.

The maternal instincts that take over in hens that become broody are impossible to predict reliably. Some breeds still carry excellent broodiness genes, such as the Silkie and heavy, docile breeds like the Cochin, Plymouth Rock, Orpington, and Rhode Island Red. Many commercial breeds of layers have been carefully selected over several years to lay continuously without turning broody, and some of the lighter, more flighty breeds like Leghorns and Minorcas are not known for their broodiness.

A broody hen needs to incubate her clutch of eggs for 21 days before they hatch. The eggs are nestled under her against a featherless patch of skin on her breast called a *brood patch*. Normally a feathered area on her body, the hormones of broodiness cause her to loose feathers on her belly, creating direct contact with the skin and acting to incubate the eggs. She turns the eggs several times each day so that the embryo inside the egg does not stick to the shell membrane. She will become very protective, ruffling her feathers on her back and emitting a shrill, shrieking sound if approached too closely. Broody hens will leave the nest unattended for short periods of time only in search of food and water and to eliminate waste. Care should be taken not to position water and food too closely to the nest, as it could attract attention to the nest site and will decrease her needed exercise (don't make it impossible for her to feed herself, though).

Broody hens should be disturbed as little as possible. Some poultry keepers suggest lifting the hen off her nest daily to eat, drink, and defecate. I have not experienced a broody hen incapable of regulating her needs by herself, and the constant disturbance might cause her to abandon her nest before the job is done. If you don't like where she has set up house and want to move her onto new eggs, let her brood in the spot she has chosen for at least three or four days. Move her to the new spot at night, taking care to give her some trial eggs, and see if she continues to stay broody. Make sure that her new nest is far enough away from the spot she originally chose, or she may return to it after a trip for food and water. After a couple of days, you can substitute the new eggs that you intend to hatch.

The number of eggs to set under the broody hen will vary by breed. A Plymouth Rock can easily incubate ten eggs, but a Cochin bantam may only be able to incubate three large-breed eggs or five to seven bantam eggs. The fact that the hen will now end up setting for more than 21 days in total won't be a problem; I once had a hen that sat on a mudpie for two months before I finally intervened.

ABOVE: Pistol, an Old English Game bantam, rests in the coop while one of her chicks scales her back. Karl Hanson gets fertile eggs from a nearby farm to be hatched out by the family's brooding hens. "I enjoy seeing what comes out," Hanson said.

ABOVE: Autumn, a Bantam Partridge Rock hen, walks with two chicks she hatched from fertile eggs brought in from a nearby farm by Karl Hanson. Hanson and his fianceé, Cloë Milek, have been raising chickens in their 2/3 acre in-town yard for the past seven years.

It is more important to be sure that a hen that is broody will stay interested in setting on the eggs until they hatch.

Only hens in good physical condition should be allowed to carry out incubation of a clutch. The energy demands required can threaten the life of the hen if she is in poor condition and decrease the likelihood that her chicks will reach maturity.

Breeds that are good brooders and mothers are not necessarily top egg or meat producers. Australorps, Brahmas, Old English Game Hens, Sussexes, Cochins, Frizzles, and Silkies make excellent broody hens and are often kept as surrogate mothers with breeds that are considered non-setters.

Broody hens typically need to be separated from other layers, as they will readily accept new eggs such that hatch dates will become unsynchronized.

Broody hens need to be safe from predators, and they tend to "hide" their nests (keep them concealed from other hens, predators, and the flockkeeper). If one of your hens goes missing, try to catch her on her trip to food and water, then follow her back to her nest.

Inexperienced setters will sometimes abandon the nest before the eggs hatch, or set on the eggs to try and hatch every last one, even if they're infertile, causing the demise of the first hatched. Broody hens need to be closely supervised for maximum success of the clutch.

After the eggs have hatched, the brood should be moved to an area free of predators and competition with other poultry for food and water. Chicks should be supplied with chick starter, but access to plenty of fresh, natural food will greatly add to the health of the chicks and the mother (see Chapters Three and Four).

PRODUCING FERTILE EGGS

If you've just decided to let the hen that has become broody hatch her own eggs, you won't need to worry about selecting eggs for her to hatch. If your goal is to maintain a particular breed or improve your flock's quality, however, you should consider which eggs to set under your broody hen.

Don't be tempted to select eggs for incubation just because they look pretty. Above all else, the most important consideration for selecting the eggs you use to hatch should be the breeding stock. Because fertility is an inherited quality in the individual hen, the hatchability of the egg can be improved by selecting the most fertile hens—those that are healthy, young, lay uniformly shaped eggs, and are observed being mated by the rooster. You should keep no more than 11 hens per rooster to ensure fertility, and the flock should be observed daily. Chickens are not monogamous, though roosters will pick favorite hens, and some individuals within the flock may not lay fertilized eggs.

Stock used to produce fertile eggs should be given a diet called a *breeder ration* that is fortified with vitamins and minerals (see Chapter Four). Ideally, access to the outdoors and insects, worms, and greens will greatly benefit the birds.

Chickens don't have external sexual organs. Instead, the hen and rooster have the same anatomy, called a *cloaca*, where the sperm is transferred from the rooster to the hen in what is called a *cloacal kiss*. A sure sign that the rooster has been mating a hen is

ABOVE: A "Barnyard Classic" chick, one of mixed heritage, ducks under a tarp in the barn. The chick was hatched by a broody Silkie hen in the farm's barn.

broken or missing feathers on her back and on the back of her neck. This happens when the rooster holds onto her neck with his beak and jumps on her back, referred to as *treading*. Treading causes the hen to drop her neck to the ground, lifting her tail for mating. Sperm can last inside the hen's reproductive tract for up to 20 days, although its viability drops sharply after 7 to 10 days. Chickens will mate throughout the year, anytime during the day, but some poultry experts claim that mating is more frequent in the afternoon, when there is less of a chance that an egg is obstructing the way for sperm to reach the mature ova.

Breeding birds should have access to at least 14 hours of light daily. Light stimulates the pituitary gland to secrete the hormones responsible for making a hen lay her eggs. Light also increases the semen output in the rooster. An inexpensive timer from the hardware store can be set to provide light prior to sunset and again before sunrise. In the northern United States, our layers are given artificial light beginning in mid-October through April.

ABOVE:
Boyfriend, a "Barnyard Classic" mixed heritage rooster, looks out the barn window. The farm's roosters fertilize eggs that are hatched by hens or an incubator.

Don't forget to check the timer after power outages, use energy-efficient bulbs, and reset the timer after daylight savings begins and ends.

Eggs from pullets and old hens should not be incubated. Porous shells (where light leaks through tiny holes in the egg upon candling, revealing differences in the shell's thickness) and abnormally large, small, or misshapen eggs should all be avoided. Cracked eggs will seldom hatch. If you are new to this, choose eggs that can be easily candled (white or lightly tinted).

The hatching eggs that you choose should be clean, although don't wash them as this can re-move the *cuticle* or *bloom* that protects the shell, and can allow more rapid evaporation within the egg. Store the eggs between 50 and 68 degrees Fahrenheit in egg cartons or flats, with the small end of the egg pointed down. Keeping the eggs for two to three days prior to incubating is ideal, but if they are held longer than seven to ten days, they are less likely to hatch. This is because eggs have genetically adapted to a hen's habit of leaving the nest after each egg has been laid so that develop-ment is arrested until the clutch is complete and the hen begins incubating full-time. Keeping the eggs longer than ten days in an artificial environ-ment such as an egg carton will cause a decrease in hatchability. There is no need to turn the eggs prior to incubation unless you're keeping them for longer periods of time.

Finally, when you are ready to begin incubating the eggs under your broody hen, place each egg under her by cupping your hand protectively over it (broody hens guard their nest aggressively, and as you slip the eggs under her, she can inadvertently peck and crack the egg; if you're concerned about getting pecked, wear a glove). If your broody hen accepts the eggs by continuing to incubate them, mark your calendar for the grand hatch in 21 days!

The Completed Egg

A hen's egg is a perfect living environment for the fertilized embryo it contains (if, in fact, roosters have been with the hens). The yolk contains the *blastoderm*, or germ that the sperm has penetrated. In the 24 hours it takes for the egg to be laid, this fertilized single cell has divided into hundreds of cells that provide the foundations for the development of organs and tissues that make up the chick. The egg contains an adequate food supply and a covering that protects the embryo during its development. It even provides the newly hatched chick with nourishment to hold it safely under its brooding mother while its siblings hatch over the next 24 hours.

CANDLING EGGS

Candling is an ancient term coined from the practice of holding the flame of a candle up to an egg to check its fertility. After 72 hours at a temperature as little as 102 degrees Fahrenheit, an intricate web of blood vessels is formed within the egg and is visible by shining a light through the shell. Eggs set under broody hens or in incubators should first be candled at a week or 10 days of age (the period at which the first critical embryo die-off occurs). White eggs are the easiest to candle; a spiderweb of vessels around a dark spot is a healthy embryo. If you see no webbing, and the egg is clear, the egg is probably infertile. Eggs should be candled a second time at about 14 to 17 days. By this time, a distinct air space at the large

BELOW:
A Golden Comet hen gets to work in a nesting box. The Golden Comet is a popular commercial egg-layer.

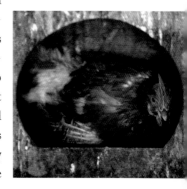

end of the egg can be seen, and the rest of the egg will be dark. If the egg looks muddy, and the edge of the air space jiggles or is uneven, the embryo has most likely died. All infertile or dead eggs should be removed as soon as possible to avoid them bursting and soiling the viable eggs. You can usually hold the egg up to your ear at about 20 days and hear the chick peeping or pecking from inside the egg.

A penlight or flashlight held up to the large end of the egg in a dark room will suffice as a candler, or you can make a device easily at home. We typically use a paper towel or toilet tissue cardboard insert. Cut the roll to about 3 inches, then place the egg, wide-end down, on the roll. Shine the light source up through the bottom. The air cell is clearly visible in an egg that has a developing embryo, and sometimes the embryo itself can be seen moving in the egg.

Using an Incubator to Hatch Eggs

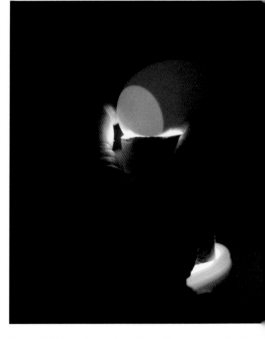

ABOVE: Using a flashlight underneath a paper towel roll, the author shines light through an infertile egg.

ABOVE: Using a flashlight underneath a paper towel roll, the author shines light through a fertilized egg—an air pocket shows up in the egg.

Incubators have several advantages over naturally hatching your chicks. First, you can time the hatch; you don't have to wait until a hen becomes broody. Second, you can hatch several more eggs at once with much less labor than it would take to hatch the eggs under broody hens that need individual attention, food, and water. Whether the broody hen will set on the eggs until they hatch is also a concern. Last, incubators are a great way to bring the joy of raising chicks into the home.

Descriptions exist of both the Chinese and the Egyptians hatching chicks by artificial

Chicken Scratch

As early as 360 BC, people have claimed that the shape or size of an egg can be used to determine the sex of the embryo inside. Aristotle, the ancient Greek philosopher, claimed that elongated eggs yielded male chicks. About 300 years later, Pliny the Elder argued that the opposite was true. In truth, about 50% of the eggs that hatch will be males—what do you plan to do with them if you are interested in a laying flock, or they are not a quick-maturing meat type? Finding a home for roosters is hard, but as science and technology grows by leaps and bounds, we may soon be able to determine the sex of chicks before the egg hatches. Until then, you should consider a plan B.

means centuries before Christ's birth, in the time of Moses, by the ancient priests in the Temple of Isis. The Egyptians used sun-dried brick structures heated by fires, where the eggs were placed on grates in heated chambers. There were no thermometers in these chambers; the attendants who lived inside the incubators could sense the correct temperature to keep the fires. As many as 90,000 eggs could be incubated at a time, for 15,000,000 to 20,000,000 hatched yearly. Eggs were tested for fertility by placing them on the palm of the hand or against the face; if the egg was cold, it was discarded. In China, smaller ovens were used. The French used wine casks, packed with horse manure and circulated hot water, over which eggs were placed to artificially incubate; and in England, hot air was passed through flues and over the eggs to hatch chicks. But it wasn't until the early 1900s that incubators could boast hatching as high a percentage of chicks as could a setting hen. Modern incubators incorporate climate-controlled heat and humidity to simulate conditions that would naturally occur. Eggs are mechani-

cally turned, and hatches are normally at least 80% successful. Incubators for home use can be purchased for small- or large-scale hatches with built-in turning devices and thermostatic controls (see Appendix Four). I prefer incubators with automatic turners, as it cuts down the time needed to open the incubator, reducing the risk of unstable temperature or humidity conditions inside.

Incubators should be placed in a location where temperature and humidity will not fluctuate wildly. Even though these conditions are regulated, the thermostat inside the incubator can't compete with drafts, temperature spikes from artificial heat sources, or unnaturally dry or wet conditions. Placing a thermometer that reads minimum and maximum temperatures in different areas of the house or barn may help choose the ideal spot for the incubator.

Typically, internal temperatures in the incubator are kept at 99.5 to 102 degrees Fahrenheit, depending on whether it is a forced-air (where a fan blows the air and circulates the heat within), a still-air, or a natural-draft (where air follows gravity within) incubator. Humidity is very important for a hatch to be successful. If too humid, the eggs will not evaporate moisture adequately, and the chicks will hatch with mushy navel disease (see Chapter Ten). Not enough humidity will cause the egg's inner shell to dry out and make it impossible for the chick to break out of the shell. I have always found this part of incubation the trickiest, and how the mother hen can pull it off without a hitch 95% of the time remains a marvel. Humidity should increase right before the hatch, and opening and closing the incubator near hatching can have marked effects on hatching success. Incuba-tors are equipped with separate water tanks that are filled to sup-ply humidity inside. The larger the surface area of the container, the more potential there is for higher humidity.

A thermometer with both a dry bulb and a wet bulb is the best way to determine relative humidity in the incubator. The wet bulb has a wick that is positioned over the end of the thermometer and placed in a reservoir of water (the bulb itself does not touch the reservoir of water). Relative humidity can be determined by finding the difference between the two readings. A dry bulb reading of 99.5 degrees Fahrenheit with a corresponding wet bulb temperature of 88 or 89 will result in a relative humidity of about 65%. This is adequate until the last three days of hatching, when the relative humidity should increase to 85% (the wet bulb should now read about 95). You can increase humidity by decreasing the airflow into the incubator (close some of the vents), or by increasing the size and surface area of the water container inside the incubator. Al-ways follow the manufacturer's instructions for your incubator and make minor adjustments depending on your conditions.

ABOVE: The author removes hatched chicks from an incubator to a brooder box.

A crude alternative to having a wet and dry bulb thermometer to determine the correct humidity inside the incubator is measuring the relative size of the air cell within the egg. This is done by candling the egg and comparing it to a standardized diagram of what the air cell should look like at a specific time of development. Humidity can be decreased or increased, depending on the size of the air cell with relation to the age of the embryo.

Setting the Eggs for Hatching in an Incubator

The eggs you choose to hatch in the incubator should be selected using similar criteria for those hatched under a broody hen. They should be similar in size, dirt and crack free, of the same age, and have the classic egg shape: wide at the top, tapering to a rounded point. The shape of the egg is important for the developing chick inside. In order to hatch unencumbered, the chick must develop with its head facing toward the wide end of the egg and the air cell. Eggs that are not uniform in size and age will prolong the hatch and increase the chance of embryo death.

ABOVE: "Barnyard Classics"—newly hatched chicks of mixed heritage—cheep and chatter while under a heat lamp.

Eggs should be placed and maintained within the incubator according to manufacturer's instructions. It's always a good idea to set up and run your incubator without the eggs in it for at least 12 to 24 hours to make sure everything is operating properly.

If you are not using an automatic turner, it is useful to make an X with a Sharpie marker on one side of the egg so you can keep track of which side it has been turned on. In the nest, eggs are turned as much as every 20 or so minutes. Manually, turning them three times a day is sufficient. Three days prior to hatching, the eggs should stop being turned. If you're using one, take the auto-matic turning device out and place the eggs on their sides in the incubator's tray. By remaining stationary on the incubator's shelf, the chick can position its beak correctly along the egg's shell. If you can, allow plenty of room between the eggs on the shelf, so that when the chicks hatch they don't knock into the other eggs.

PROBLEMS WITH THE INCUBATOR

Ideally, an uninterrupted power source should be used with the incubator. If you don't have one, the best thing to do if a power outage occurs is to mimic the conditions that are present when the setting hen leaves her eggs to eat, drink, and defecate. Open the incubator and let the eggs cool. This will also avoid the possibility of suffocation, now that the ventilation fan is not running. Usually, if power is restored within 12 hours, it will not affect the embryos. Make sure to check the temperature within the incubator after an outage, because the thermostatic wafer, filled with ether, may have been affected. It is always a good idea to keep an extra wafer and thermometer on hand when using an incubator (these can be purchased through the same supplier that you purchased your incubator from).

Ventilation within the incubator is also important. Without proper airflow, the unhatched embryos can suffocate. As the hatch approaches, the need for ventilation increases. Unfortunately, increased airflow can decrease humidity. A humidifier placed in the room that holds the incubator can help, or water reservoirs within the incubator can be increased.

Degrees Difference Between Wet and Dry Bulb

	1	2	3	4	5	6	7	8	9	10	11	12	13	14	15
60	94	89	84	78	73	68	63	58	53	48	44	39	34	30	26
65	95	90	85	80	75	70	65	61	56	52	48	44	39	35	31
70	95	90	86	81	77	72	68	64	60	55	52	48	44	40	36
Dry Bulb Temperature 75	95	91	87	82	78	74	70	66	62	58	55	51	47	44	40
80	96	92	87	83	79	75	72	68	64	61	57	54	51	47	44
85	96	92	88	84	80	77	73	70	66	63	60	56	53	50	47
90	96	92	88	85	81	78	75	71	68	65	62	59	56	53	50
95	96	93	89	86	82	79	76	72	69	66	63	60	58	55	52
100	96	93	89	86	83	80	77	73	70	68	64	62	59	56	54

RELATIVE HUMIDITY TABLE (%)
The relative humidity within the incubator can be found by reading the dry bulb temperature and then finding the corresponding difference in degrees between the dry and wet bulb. For example, a dry bulb temperature of 85% and a wet bulb temperature of 90% would be a difference of 5 degrees, resulting in a relative humidity of 82%.

ABOVE: In a cardboard box in the living room, a newly hatched chick sits under a heat lamp after it emerged from its egg in an incubator.

Stages of Hatching

As the hatch date approaches, conditions in the incubator become more critical. While the hatch is occurring, the incubator should be disturbed as little as possible to keep temperature and humidity consistent. The embryos are easily chilled, and their chances of hatching will be compromised if you open and close it too much.

Live embryos within the eggs begin vocalizing as much as 24 hours before they break free from the egg's shell. Close examination of the egg will reveal tiny cracks, radiating from a point somewhere along the widest part on the egg. This is called the *star crack*. Eventually, the crack turns into a small hole. This stage is called *pipping*. Closer investigation will reveal the tiny beak of the chick armed with its special egg-cracking tool, the *egg tooth*. The chick uses the egg tooth to make the series of breaks along the eggshell, and eventually, it uses its legs to push the small end of the egg away and bursts from the shell, wet and tired. Chicks are hatched with enough nutrients to take them through 24 to 48 hours without additional food or water. In this way, the hen can naturally and safely set on her clutch, allowing the last of the eggs laid to hatch. In the incubator, shells should be removed quickly from the hatching tray, and chicks should be placed on the bottom shelf to avoid injury or jostling of unhatched eggs. Chicks can be removed from the incubator when they are dry (about 18 to 24

❋
· · · · · · · · · · · THE PERILS OF · · · · · · · · · · ·
ARTIFICIAL HATCHING

It is day four of the hatch. Our incubator is an ancient one, and because of that, along with the drafty farmhouse conditions, our −10-degree-Fahrenheit nights, and the fact that our hens are different ages and not in their breeding prime, our hatch has dragged out longer than desirable. I have set the eggs in order to have chicks to show to a class of children before February break, so the hatching conditions are neither perfect, nor ones in which I would normally choose to set eggs. I announce to my son that it is time to pull the plug on the remaining unhatched eggs. He lets out a shriek of protest and gallops over to the incubator to stop my hand.

At eight years old, he is all boy. He breaks tree branches in mock swordsmanship; the cats are terrorized by his flying leaps onto their sleeping forms for "snuggling." He pulverizes the burdock plants along the barn's length, and rotten pumpkins are reduced to glistening mounds of orange, putrid pulp. I am apprized of flies that have succumbed to the pencil sharpener or the woodstove, and tomato hornworms become gooey globs of green that often bespeckle his clothes and face. With the exception of the hornworms, these antics receive sharp denouncements from me, his mother, who for his last hours after hatching) and kept in a brooding pen with a heat lamp that maintains a temperature of 95 degrees Fahrenheit at a height of 17 inches from the floor of the pen. If the hatch is prolonged, remove the chicks to the brooder (see Chapter Three) every 8 to 10 hours and provide them with food and water.

birthday bought him a piñata shaped like a cactus, because the thought of allowing a band of eight-year-olds to beat animal or human shapes to smithereens was reprehensible.

With the chicks, he is different. On the day we were to open the incubator and remove the dry chicks to the brooding box, he woke up 30 minutes early. On the following morning, while the newly hatched chicks waited out an Arctic front in a box in our living room, he tiptoed down the stairs to keep from waking us and gingerly held their cotton-ball forms on his lap. Each one was quickly identified according to color or peep pitch. And each was held with the dread that it would become a rooster and eventually end up in our freezer.

So now we are upon the hour, time to stop the incubator. We pull out the remaining eggs and carefully hold each to our ear, listening, searching, and hoping for signs of life. In the end, 70% of our eggs hatch. Not bad for the odd assortment of eggs that we had set. We carefully investigate each of the unhatched eggs, identifying what traits they had that might have contributed to their demise: not the right shape, too large, too small, a hairline crack that had not been noticed before. Then, we collect the remaining eggs and compost them. We unplug the incubator, and sit, delighted at the whirling dervishes in the brooder box that chase after each other's toes, beaks, and scattered pieces of grain. I hope they're all hens, he murmurs. I secretly smile, thankful for my chicken-sensitive son.

Raising Day-Old Chicks

The most popular method of starting a flock is by purchasing day-old chicks from an established hatchery. Only chicks that are healthy are shipped, avoiding the problems associated with malfunctioning incubators and inexperienced broody hens that don't tend to their chicks. Often, chicks are offered as sexed rather than straight-run (these are unsexed chicks that are more than likely 50% males). This option is especially nice if you'd rather not bother with dispensing with the cockerels once they've grown. Remember, free roosters are not exactly a sought-after luxury item.

Chicks are shipped the day of the hatch and arrive within 48 hours by U.S. Postal Service. Mark your calendar the day before expected delivery, so that the brooder is ready for the new arrivals, and you don't find yourself hunting for the heat lamp and waterers when the post office calls you.

In rural areas, the USPS is no stranger to the arrival of chicks in the mail. Our local postmistress, Gloria, calls me the minute they arrive in our tiny town, regardless of the early-morning hour. Even locked, the secret chick knock on their front door brings the box of peeping babies out to the lobby. Visit your post office and warn them of the chick's imminent arrival, and give them reliable phone numbers for where to reach you. If your central distribution center is close, as ours is, you could even ask them to phone you. It may mean the difference between chicks spending the night just 25 miles away because they've missed the ride before the close of your rural post office, and most postal workers would just as soon have the noisy things taken away. Make sure to inspect your package right there in the

Chicken Scratch

Don't feed the shells of hatched chicks to your chickens. Throw them in the compost. Be sure that children handling the chicks and their hatched shells wash their hands thoroughly after handling them. They are loaded with bacteria and feces from the embryo.

If all the conditions are maintained correctly in the incubator, hatches of up to 85% can be achieved. A normal hatch takes about 24 hours.

LEFT: Bradford Jones, 9, gets an up-close view of a newly hatched chick that was hatched by incubator.

	Problems with Eggs or Breeding Stock	Problems with Care and Handling of Eggs	Problems with Incubation
Infertile	Too few males Too many males Males too old Preferential mating Hens too old, too thin, too fat Stress high in breeding pen	Eggs too old Eggs improperly stored small-end up Eggs held at too high or too low temperature Eggs held at too low humidity	
Fertile, but Embryos Die at Early Age	Eggs too old Insufficient nutrients fed to breeding stock (lack of vitamins A, D, E) Hens too old	Eggs too old Eggs improperly turned Roughly handled eggs	Incubator temperature unstable Humidity too high or too low Insufficient turning of eggs Eggs placed incorrectly in incubator
Die at 12 to 21 Days (Unpipped)	Breeding stock fed insufficiently Unclean or cracked eggs	Improper sanitation	Poor ventilation Insufficient turning of eggs Incubator temperature incorrect
Die at 21 Days (Pipped)	Unclean eggs Paratyphoid	Turning eggs past day 18 (malposition of embryos)	Humidity too low Poor ventilation
Unable to Hatch Out of Shell			Temperature too high Humidity too low Too much ventilation
Splayed Legs			Temperature too high
Crooked Toes			Temperature too low
Unabsorbed Yolk Sac			Humidity too high
Mushy, Smell Bad			Unsanitary conditions in incubator
Crossed Beak	Genetic, cull chicks		

ABOVE: The author picked up a mixture of Bard Silver cross chicks and Plymouth Rock cross chicks from the South Royalton Post Office. The farm raises batches of 100 to 150 chicks every six weeks in the summertime to sell to customers for their meat.

Lucy, a three-month-old Buff Orpington/
Plymouth Rock cross pullet, roams the
backyard pen at the home of Geoff Hansen
and Nicola Smith. Hansen found the ten
pullets online and drove across the state to
buy them for himself and a nearby friend.

ABOVE: A Cuckoo Marans hen, top, and an Ameraucana cross hen get a view of the morning from a barn window. The farm has 50 laying hens, producing two to four dozen eggs a day.

lobby, and if need be, fill out a claim form for any dead chicks (most hatcheries have specific requirements that need to be met in order to qualify for reimbursement).

Started Chickens

For anyone considering keeping chickens primarily as a laying flock, partially grown birds may be the best option. Although they don't tend to be as tame as chicks that have grown up in your flock, they are more economical in the long run. Hens don't typically begin laying eggs until at least 20 weeks of age, and some of the heritage or dual-purpose breeds may take six or seven months before laying. During this time the average chicken bred for commercial production of eggs will eat 15 pounds of feed, while the slower-maturing, heavier-bodied birds may eat as much as twice that amount. Mortality is not as high in started birds as it is in chicks you've hatched or purchased.

Be sure to buy started birds from reputable breeders who know the importance of feeding young laying hens correctly and who maintain well-kept, disease-free flocks.

Free Chickens

The papers are full of free chickens, mostly roosters, but sometimes layers, always claiming to be just a year old. As enticing as they may seem, try and avoid free chickens unless you know the person giving them up. First, chickens are rarely just a year old if they are being given away. More important is the risk of introducing some sort of disease into your chicken yard. Otherwise healthy-looking chickens can harbor anything from lice to scaly leg mite (see Chapter Ten). Usually, unless there has been a tragedy in the family, the reason to unload chickens is because they no longer serve a useful purpose to their owner.

One of Fat Rooster Farm's hens takes a look at newly delivered chicks that had been placed in a brooding box. When the lid is closed, a heat lamp keeps the chicks warm as they adjust to their new surroundings.

(3)

Housing Chickens

The chicks have come a day early. I'm usually only half prepared for their arrival by mail anyway, always busy trying to find the three-prong adapter for the extension cord or marbles to stick in the water tray so the chicks won't drown. Today is even more complicated, because I don't have the day off, and I have an early-morning meeting to make. To make things worse, I have the flu, and don't feel like doing anything at all.

Typically, when Gloria from the post office makes the 6 AM call to the farm, I'll rush down to collect the boxes of babies, then spend the whole morning checking their feed, their heat lamp, and their water in the brooder box.

Instead, today I hastily prepare the box, install the heat lamp, fill the waterer, and sprinkle chick starter mash on the newspaper floor. The box is only a temporary house. It acts as an intermediary between their shipping box and the pen where they will spend the first few weeks of life. From this pen, they're moved to an outside coop, where they'll forage for insects and eat vegetation in addition to the grain. I leave the chicks to attend my meeting.

The brooder box has a finicky light switch. If a hen decides to roost on the outside of the box and hits the cord just right, it snaps off the heat lamp inside, leaving the chicks in total darkness and cold. Usually the light will snap back on after the chicken has jumped off the box, and the chicks will resume scurrying across the newspaper, picking up grain, and chasing each other.

But when I return to check on the box three hours later, there is chaos inside. The heat lamp is out, and at first I really can't tell what I'm looking at. The little fluffy bodies are now soaking wet, and they are lying still, stretched out in unnatural positions on the soggy newspaper. Apparently all 50 white chicks have drowned in the waterer, piling one on top of another, after the heat lamp had gone out. Why they have decided to commit mass suicide in the waterer is beyond me.

The sheer waste of life that I could have prevented overwhelms me. I sit by the box and start to cry. Sobbing so hard in fact, that the little black-and-white barn cat, which rarely shows his face, crawls over to me and swishes his tail, back and forth, puzzled at what I am doing. I stop the tears to pick him up and stare at the chicks again. Then I notice that some of them are moving, their bodies twitching and fighting for life.

I jump up, spilling the cat to the ground, and scoop the moving chicks into my tee shirt. I run to the house, grab a baking dish, and pop them in it. Then I set the gas oven to 110 degrees Fahrenheit and stick them in. Within minutes, I hear the happy sound of warmed chicks, in search of food. I keep thinking to myself that it's time to change that fixture on the brooder box as I cheerfully scoop up the chicks that are "done" back out of the baking dish. This time, I was lucky.

LEFT: Cornish Rock cross meat bird chicks warm themselves in the early morning.

Housing Your Chickens

When we began raising meat chickens, they lived for the first few weeks in the room we grandly call the library, next to the computer. After the ventilation fan failed in the motherboard, we began raising them in the unfinished bathroom. Their dander and the grain dust turned the ceiling brown, so that when anyone ran the hot water in the sink, condensation would form on it and brown drops slowly fell to the floor. From that day forward, we decided that chicks should be raised where furniture and equipment could be spared—in a chicken house.

Newly hatched chicks need a draft-free area that is warm and dry. This can be as simple as a cardboard box lined with newspaper and a bare lightbulb for heat, or as fancy as an entire area of your coop dedicated to new chicks.

What you are attempting to duplicate are conditions that the mother hen would supply to her chicks. Normally, the hen will cover her chicks, or "brood" them, providing shelter and warmth.

ABOVE: Doing what a mother hen would usually do, the author introduces newly delivered chicks to water in a brooding box. Stones in the water dish prevent them from falling into the dish and drowning.

She listens for cues to determine when they are hungry, and then she will unceremoniously rise and lead them to food and water when necessary.

In their artificial environment, chicks should have enough space to be warm but not overheated. Overcrowding can create sanitation problems, suffocation, and even cannibalism. Ideally, the pen that chicks are started in should have rounded sides, so that there are no corners for chicks to pile up in. From day one to three to four weeks old, each chick will need about half a square foot of space. From age four weeks to 12 weeks, the amount of space required for each will increase by a square foot every month. Full-grown chickens need about 3 square feet each.

The temperature in your new chicks' pen should be about 95 degrees Fahrenheit for the first week of life. Heat lamps with 250-watt brooder bulbs placed at least 17 inches from the bottom of the pen will adequately warm a 12-square-foot area in an 8-foot by 4-foot by 3-foot-high pen, provided that it is draft free. A 100-watt bulb in a cardboard box will provide heat just as well if you are raising the chicks in your kitchen for the first week of their life. Put a thermometer on the bottom of the pen below the bulb and check that it registers 95 degrees. Any hotter, and you run the risk of the bedding catching fire; brooder lamps are fitted with reflectors and a wire guard so that if they do fall, there is less chance of paper or chicks becoming burned. The temperature within the brooder can be decreased by 5 degrees every week until it reaches 70 degrees. At this point, the chicks will be about three weeks old and capable of maintaining their body temperatures adequately, provided they still have adequate shelter from cold, wet, or extremely hot conditions.

Observe your chicks in their new environment. Contented chicks will emit soft cooing calls, not shrill panicked sounds. They will be scurrying about the pen, searching for food and drinking occasionally. If they're all huddled under the heat source, they're most likely too cold. If they are hugging the sides of the pen, avoiding the heat lamp, they are probably too warm.

At Fat Rooster Farm, we use this succession of pens before the chickens are turned loose on pasture:

DAY ONE TO DAY FOUR OR FIVE. Chicks are removed from their shipping boxes to a plywood brooder that measures 2 ½ feet wide, 21 inches high, and 3 ½ feet long. The bottom of the box is mesh hardware wire, lined with newspaper (the non-glossy type). The waterer is put at the far end of the box, away from a heat lamp that hangs about 17 inches from the bottom of the pen. The heat lamp is a red-tinted 250-watt bulb in a porcelain fixture. The chicks are provided with light 24 hours a day to encourage them to eat and keep them from panicking and suffocating if they pile on top of one another for warmth. The top of the pen is hinged, and propped open for ventilation. Pebbles or marbles are placed in the water tray to discourage the new chicks from piling into it and drowning. Grain is sprinkled on the newspaper for the chicks during these first few days. Each of the chicks has its beak dipped into the waterer to ensure that it knows where to drink—remember, the mother hen would cluck and call her chicks to food and water. She would drink and pick up grain to show them what to eat. While chicks can instinctively feed and drink, teaching them how to do these tasks will decrease their chances of becoming weak and dying after their stressful journey to their new home.

DAY FIVE OR SIX TO THREE WEEKS OF AGE. The chicks are moved to a "chicken tractor." It's referred to as a tractor, because it's a pen that is four-sided, with a hinged top, but with the base of the pen being entirely open. This allows the chickens to scratch up and

LEFT: Cornish Rock cross meat birds huddle together in the barn at Fat Rooster Farm. Due to customer demand, the farm raises batches of 100 to 150 meat birds every six weeks through the summer.

till the soil, like a tractor would, fertilizing and cultivating it in one fell swoop.

There are several versions of tractors available; the one that we have settled on is lightweight and easy to construct. Our pens, made from two-by-fours that have been ripped down the middle, measure approximately 8 feet by 4 feet by 28 inches high and have a total area of approximately 32 square feet. Each side and the top is constructed separately and is held together with carriage bolts so that the pens can be dismantled at the end of the season and stored flat against a wall. The corners of each of the panels are braced with wood, and the sides and tops are covered in 1-inch chicken wire. The top is covered with waterproof material like a plastic tarp, and three of the pen's sides are covered with plastic or tarps. The tractors are set up in a sheltered structure (in our case a converted dairy barn) where the temperatures don't fluctuate wildly between night and day, and where wind and rain are not a threat. The pens are four-sided with a hinged top, completely open on the bottom, and the floor of the barn is spread with pine shavings. Don't use cedar shavings—they can irritate the chicks' eyes. Using hay can create obstacles for the chicks as they attempt to walk around, and it is not very absorbent or easy to clean up afterwards.

For every 50 chicks there is a 1-gallon waterer placed within 2 feet of the heat source. Grain is put in trough feeders placed within 2 feet of the heat source. During the first two weeks, there is a 250-watt heat lamp in either corner of the pen.

······Moving Chicks Outside·······

Transitioning chicks from a pen that has been heated artificially, kept from wind, rain, and sun, and kept free of predators to pasture can be difficult. Before the move is made, you should:

- Wait until the chicks have developed their *scapular feathers*, the little band of feathers above their wings on their backs. The feathers look a little like the shoulder pads on football players. They act to insulate the chicks from sun, cold, or wet conditions. Typically, chicks develop these feathers at about three weeks of age. Chicks should not be moved outside if nighttime temperatures within the pen can't be regulated above 60 degrees Fahrenheit, otherwise, they'll spend most of their energy keeping warm rather than growing.

- Wait until weather patterns are stable for your move. Choose a day to transition the chicks when the next few days don't threaten steady rain, thunderstorms, heavy winds, or intense heat.

- Move the chicks well before sunset so they can grow accustomed to their new surroundings. Put familiar watering and feeding devices in their pens to encourage them to eat and drink immediately.

- Have at least three sides of their new pen enclosed, and provide a heat light on a timer for the first few days to provide them with heat and light at dusk through dawn. Red-colored bulbs will disrupt their sleep patterns the least and may attract fewer predators than the white bulbs. At night, the pens should be predator proof, either closed tightly, or surrounded by electrified poultry fence (Appendix Four).

- Monitor the chicks closely through the first few hours in their new home. They should be scattered about, busy foraging, emitting few, if any distress sounds (high-pitched, constant peeping sounds instead of pleasant, cheerful notes of content).

- Don't mix different batches of chicks together at the time of the move. This can lead to bullying or distraction from eating and drinking.

Silver meat birds
huddle together
in a chicken
tractor—a
portable coop.
The coop allows
for the birds to be
on fresh grass and
distributes their
manure.

- Make sure that feeding and watering stations are sheltered so that the young birds are not forced out in downpours or extreme heat to forage. These can either be areas inside an enclosed pen that are covered, or individual range shelters that act to keep feeding stations dry. Changing the location of feeding and watering stations will also distribute areas littered with grain and wet ground. The birds will tend to defecate more near where they are fed and watered, so manure will be more widely distributed if the stations are moved daily.

- As the birds mature, your management of feeding areas and shelter will change. When the birds are small, their feeding stations may need to be changed less frequently. As they mature, you may find yourself moving feeding stations and pens every day where they seek outshelter during inclement weather and at night. You may need to rotate the electrified boundary fences to change out their yards. At Fat Rooster Farm, we start out each new batch of chicks that is introduced to pasture by providing an enclosed shelter surrounded by electrified poultry fencing. As the birds grow accustomed to their outside quarters, we remove the fencing, allowing them to mix with the other birds. There is still predator protection along the perimeter of the common yard, and the night shelters continue to be moved so that a manure pack doesn't build up where the birds roost during the night.

ABOVE: A Plymouth Rock cross cockerel meat bird rests on a log in the pasture.

Just because you have observed an animal that could potentially prey upon your birds does not mean that you should break out the AK-47 ... yet. Eating livestock is a learned behavior for wildlife, and many individuals will never be attracted to the barnyard or connect your chickens with the dinner bell. If you go out and remove a potential predator when it is not really harming your birds, you also run the risk of another individual coming in to replace the one you've removed that does know how good free-range chicken tastes.

The type of predators that may harm your chickens will vary depending on your location. However, the big seven will most likely be: hawks, owls, foxes, coyotes, raccoons, skunks, and weasels. Snakes, domestic dogs, feral cats, rats and other rodents, and anything from a mongoose to a mountain lion could also be attracted to your chicken buffet. Determining what kind of predator you have can help you catch and kill, or at least deter further losses.

Hawks and Owls

All birds of prey are federally protected, and purposely harming one in any way carries steep fines. Don't set out to kill a hawk or owl unless you're prepared to face the legal ramifications.

One winter, we had a barred owl who repeatedly found his way inside the chicken house and killed three of my favorite hens. We contacted the wildlife experts in the area, and they relocated the owl for us.

Chickens that have been killed by birds of prey are usually left behind rather than carried off. Often, they are headless, their breasts are plucked of feathers, and the meat as well as the innards have been consumed. If you examine the carcass closely, you may see talon marks where the bird has held it while dining. Other signs that a raptor is responsible for the losses are impressions of wing feathers in the snow or dust, or an occasional lost feather that is clearly not from your chickens.

RIGHT: A barred owl roosts in a tree while visiting a village backyard in Wilder, Vermont. While they are a nuisance toward chickens, all birds of prey are federally protected and require contacting wildlife experts for assistance in relocating them.

Pastured poultry should be provided with overhead shelter of some sort, whether it is an enclosed box that is moved daily, bird netting over their foraging yard, or even a truck cap that they can duck into if a bird of prey appears. Most hawks will prey upon chickens during early morning or evening, so confining the chickens to a protected area at these times might prevent attracting hawks to the area. Owls, on the other hand, like to hunt after dark, so penning the chickens up will avoid losses. Setting your birds up near outbuildings and areas with human activity can also deter winged predators. Avoid pasturing the chickens near woodland edges or dead trees that can be used as perches from which hawks or owls can study potential prey.

Foxes and Coyotes

Foxes and coyotes are legally hunted during the year in most states. Check with your local wildlife officials for specific rules

pertaining to your area. These animals can also be removed outside of hunting season in many states if you obtain a nuisance wildlife permit.

Both foxes and coyotes can do a lot of damage in a short time. In fact, foxes will even kill and partially bury their prey in frenzy, attempting to stockpile as many as they can. Just as often, however, birds will just go missing, having been carried back to the den, with little or no trace. Both tend to hunt in the early morning or evening, although hungry individuals can prey upon your birds at any time if they are desperate enough.

Again, confining the birds at night and in the early morning will make them less attractive; some say that domestic dog hair scattered around the chicken yard or human urine is also a deterrent. Playing a radio near the coop may also help. One friend uses a guard dog in his chicken yard, much like a sheepherder would.

LEFT: Whitey, a White Rock, wanders amongst the bushes at Carol Steingress's and Rick Schluntz's home. The couple have been keeping 5 to 6 laying hens in their small in-town yard for the past 10 years. "I think they're so interesting and curious," Steingress says. "We're hooked."

CENTER: With its prey caught, a red fox looks to return to its den on a warm Spring morning in Strafford, Vermont.

RIGHT: Willow, the family dog, explores the manure left behind by chickens outside the coop at Karl Hanson and Cloë Milek's home. The coop used to be a shed attached to the back of the garage; Hanson proposed to Milek on Valentine's Day when he left a ring for her in one of the chicken's nesting boxes.

Raccoons and Skunks

Raccoons can also be removed if they prove to be "nuisance wildlife." Again, contact your local wildlife officials for specific laws protecting these animals before doing something that will get you in trouble later.

Raccoons are extremely crafty. They are capable of pulling staples, opening windows, unlatching doors, and practically serving up the chickens, which you are so desperately trying to protect, with a side of pasta and marinara sauce. They, too, can wipe out a flock of chickens in short order. We had one raccoon that would lure the chicks to the chicken wire so that it could reach in and deftly pluck them from the cage by their necks. Another stopped raiding the coop while the traps were set and resumed destruction the day they were removed.

Raccoons tend to kill and eat their prey at the scene of the crime. Usually, there are sharp punctures near the bird's head, and the body will be ripped open and partially eaten.

BELOW: Goldie, a three-month-old Buff Orpington/Barred Rock cross pullet, forages in the backyard pen with her Black Sex-Link coopmates at the home of Geoff Hansen and Nicola Smith. Hansen decided to use an electric netted fence to keep the family's two beagles a safe distance from the birds.

The easiest way to catch a raccoon is to hide a live trap such as a Havahart near your birds. I use hay bales on either side and on the top of the cage that creates a tunnel into the trap. Salty bait like ham, mackerel, or leftover chicken work well as enticements into the trap.

Skunks can mimic a raccoon in their method of killing and eating chickens. Generally, though, their odor at the scene of the crime is a dead giveaway. They are also not as fast or as crafty at catching chickens, so you may find living birds with tiny puncture wounds near their heads or tails. Because it is harder for them to catch the chickens, they'll usually devour one more fully, say, than a raccoon will, which might leave the intestines and other organs in its haste to find another victim. Skunks can be live-trapped, and provided that they are dealt with gingerly, there is little risk of being sprayed during their dispatch. They do tend to spray after being shot, though, so if you choose this method of capture, shoot upwind.

Predators

Last night, while I was patrolling the meat birds with the gun, the air was filled with fireflies. I know that we "self-sufficient" folk are supposed to eschew modern gadgets, like iPods and such, but I can't describe the feeling of walking through the whirling flashes of lightning bugs, listening to Conor Oberst sing about his inability to hail a taxi in the city. His frustration with what seems so perfect in the evening turning to empty baggage in the morning was a fitting juxtaposition to my hectic day: fireflies making a light show for me by night, when everything had been put to bed, when all is quiet and magic and mist, waiting for that raccoon to show up, petting the barn cat milling around my feet, remembering how lucky I am to have this place.

Weasels

By far, I am most afraid of weasels when it comes to predators of poultry. They are capable of killing an entire flock of 20 or 30 chickens in a night, and they are extremely hard to capture. Weasels tend to eat just the heads of their prey first, so if they are overcome with a feeding frenzy, you can expect to find many dead chickens with nothing but their heads missing. A bird of prey, on the other hand, will very rarely kill more than one or two birds at a time. Add

ABOVE: Rosie, a Black Sex-Link hen, wanders back to the entry of a backyard coop. Designed by his friend Alex Cherington of Hartford, Vermont, Hansen and Cherington built the "Mini Coop" together over four nights and weekends.

to this the near impossibility of capturing a marauding weasel, and you'll understand my fear.

Weasel-proofing a chicken house requires using fine mesh wire such as hardware cloth. Generally, they hunt at night, so confining your chickens to a "safe house" will deter these predators. In the past, I have opted to move the chickens' outside locations to a different site rather than attempt to foil a weasel; if you have the space, that may be the easiest course of action to take. Using rat traps that have been tethered to the ground can also be effective.

Domestic Animals

Dogs and cats are also natural predators of chickens. Although their need to kill for food may be lacking, their predator instinct is often times still intact.

My clumsy, sweet, 3 year-old black Labrador retriever, Dart, will mock-chase his quarry. He'll scatter the chickens, turkeys, and peacocks around our yard like leaves falling from a tree in an early Nor'easter. The problem arises when a companion joins in the fun, like my sister's boxer, Edith. Then, a competition begins, and whoever can make the chickens more frantic becomes the victor, sometimes at the expense of the chicken.

When you make the decision to allow new dogs near your flock, be aware of the possibility of "pack mentality," like Dart and Edith's. Even if your acquaintance assures you that her dog is fine around birds, the game may change when a rival player enters the field.

Domestic dogs very rarely carry off what they've killed. Usually, birds are left in the yard or near the coop, badly maimed or injured. As a flock owner, you do have legal rights regarding the protection of your flock from domestically owned dogs. It's always a good idea to get a town or city official involved in wading through this situation; confronting a neighbor after such a devastating event can be very emotionally charged and may not result in the best outcome.

Domestic cats can also pose a threat, particularly feral cats. They are responsible for millions of songbird and rodent deaths— including deer mice, shrews, voles, and grounds squirrels, not just

BELOW: While house cats pose far less threat as predators to chickens, domestic dogs, foxes, raccoons, and skunks as well as weasels, minks, and birds of prey can wreak havoc on a flock. Having a chicken coop is somewhat akin to hanging up a "Free Chicken Dinner" sign, unless you protect your flock.

the dreaded Norway rat. Back yard coops in urban areas are easy targets; to protect against the threat to your flock, a safe yard to roam and a secure roost at night is needed.

Permanent Quarters

Chicken coop designs can be elaborate affairs, ranging from two-story buildings several square feet in area to mini-coops that house enough chickens to supply a family with eggs. Our laying and breeding stock winters in a converted wing of the dairy barn that was used to raise the young cows (heifers). It has windows on three of the four sides, a cement floor, and doors on either end.

There are several websites that contain plans for coops, and most extension agents have floor plans for coops (see Appendix Three and Four). Regardless of which design you choose, there are several points to consider when you decide to build your coop:

- It should be easily entered to clean. Make sure the doors wide enough for a wheelbarrow and there no obstacles in the way. You don't want anything to make you put off cleaning efforts, so make it easy to accomplish the task.
- The coop should not have extremes of temperature. It should be free of drafts and provide shelter from rain, wind, and extreme heat or cold and stay well ventilated.
- Ideally, electricity and plumbing should be close by so that feeding and watering is a simple task. In the winter, when daylight diminishes, it is very handy to have an outlet for an artificial light source that can be used to stimulate continued egg-laying.
- There should be roosts to encourage sleeping within the coop and nesting boxes to entice hens to lay their eggs indoors rather than in the flowerbeds or the hay pile.
- The coop should be large enough to house your birds comfortably; having a separate area for new arrivals to the flock, sick, injured, or breeding birds is also handy.

ABOVE: After Erin Regan lost her first flock of laying hens to a fox, she built a second-floor coop over her Bethel, Vermont, sheep barn. With its "free-range on-ramp," the coop's door is opened with a rope on a pulley. Regan also plays a radio around-the-clock to keep the predators away.

Location

Make sure to locate your coop in an area where there is good drainage, adequate light, and good airflow for ventilation. A south-facing exposure will provide good light and solar gain during the colder months. Having the coop too high on a hill can expose the birds to excess winds in the winter and greater temperature extremes in the summer months. On the other hand, placing it in a hollow or depressed area will make it harder to access during wet periods and will create muddy, unsanitary conditions.

Cross-ventilation is important for good health, as it removes ammonia and excess moisture from the air (it also comes in handy when you need to clean the coop and don't care to asphyxiate). If your coop includes a fenced-in yard, be sure to locate it away from snowmelt or flash-flood potential. Ideally, the yard should lean more toward sandy loam rather than clay so that precipitation quickly percolates into the ground. If you have heavy soil, you

ABOVE: Jennifer Hauck gives her Plymouth Rock pullets a treat of spoiled yogurt in the pen outside the "Mini Coop" designed by her partner, Alex Cherington. Cherington believes the small design of the building will help the chickens stay warmer in the wintertime.

might consider bringing in sand to the yard area to facilitate good drainage.

Design

The interior layout of your coop should provide roosting, feeding, and egg-laying areas (if you are keeping laying hens). Enough space for your flock to exercise and dust-bathe when it spends time inside the coop will prevent overcrowding that can lead to cannibalism. A general rule of thumb is 3 square feet per bird, but this can vary depending on how much time your birds will be allowed to spend free-ranging, the breed of bird you choose to raise (are they flighty or docile, heavy-bodied or bantams?), and your climate (how many inclement months during the year will the birds prefer to spend inside the coop?).

What materials you use to construct the coop will depend largely on the type of enterprise you choose. Obviously a concrete floor is not what you'd want for a portable coop, nor is it ideal for meat birds that you intend to free-range. Budget constraints could also determine construction design.

Windows

Windows provide light and ventilation, both of which are crucial for the health of your birds. The most practical windows should be able to open completely during hotter months when maximum air circulation is needed to keep your birds cool. They should also be able to close tightly during winter months to keep your birds warm. We usually cover the window glass with a layer of plastic during the coldest months to cut down on drafts.

Having ventilation holes at the top of your chicken house will serve as a means of ventilating the coop when the windows are shut tightly for the winter. The holes can be rectangular openings located along the roof's eaves or circular holes located in a series along the top of the coop's walls. Be sure to have a way to seal the ventilation holes up on the coldest nights.

Some poultry keepers build a sun porch on the south side of their chicken coop that provides them with a warm, comfortable space and enough light to keep them laying through most of the winter. The sides of the porch can be banked with plastic, leaves, or bales of hay to cut

BELOW: Farmer Ray Williams unhitches the tractor after moving the portable chicken coop at Back Beyond Farm. Williams and his wife, Liz York, switched to using a portable coop after the free-range chickens began to get into the greenhouses and were pecking at the tomatoes they raise to sell.

out drafts, and the entrance to the porch can be open and shut at night to reduce the area in the coop exposed to outside temperatures.

Doors

The chicken's door to the outside should be separate from your access to the coop. The chicken's door can be just over a foot high and wide, just big enough for them to come and go. This door can be hinged at the bottom and latched at the top so that it can be closed at night and during inclement weather. Your access to the coop should be larger. Remember to make it so that cleaning out the coop can be done easily. If you have different rooms in your coop, like we do at Fat Rooster, there should be chicken and human entryways to each. Old storm doors work very well in chicken coops. They also have the advantage of being able to be fitted with a screen during warm weather for added air circulation.

Floor

Cement floors reduce the number of rodents that will be able to access the inside of the coop. They are easy to shovel out and wash down. At Fat Rooster Farm, we have a soil-covered area in the concrete where the birds can dust-bathe during the winter months.

Wood flooring will also work, but the manure will eventually rot the wood, and moisture from watering stations and manure can create a slippery mess. Another disadvantage is that rodents can chew through the wood, making holes in the floor.

A wire floor works well in permanent or movable coops. It also has the advantage of effectively removing manure from where the birds stand without having to physically haul it out of a cramped space with a wheelbarrow and shovel.

A dirt floor is the most primitive choice and the easiest. A dirt floor can quickly turn to mud, though, and it is the hardest of your options to keep clean and free of disease.

Walls

The walls of the coop can be insulated to reduce temperature extremes; our coop has a plywood layer between the outer boards

ABOVE: An open-design of the portable coop used at Back Beyond Farm allows the rooster and hens to come and go as they please; a portable electric netted fence keeps predators at bay.

to create a dead-air space. The walls are smooth and can be easily whitewashed and wiped down. If you insulate the walls with Styrofoam or insulation board, be sure that it is covered with a layer of sealer or wood; otherwise chickens will peck and eat the foam if they become bored.

Nesting Areas

Egg-laying areas within the coop can be as simple as empty plastic milk crates suspended above the floor or as sophisticated as commercially developed metal trap nests hung along the coop's wall. The point of a nesting box is to create a dark, quiet space where a hen can sit to lay her egg, free from the traffic of birds coming and going inside the coop. It's also important to make a nesting area that is easily accessible for collecting the eggs and replacing the nesting material within.

General measurements should be 12 inches deep by 12 inches wide and 12 inches high. If the nesting boxes are enclosed,

A hen looks out onto the pasture from the top of the portable coop's nesting boxes at Back Beyond Farm.

the roof should be steeply pitched to discourage birds from roosting on it at night where they will create a manure pile near the clean eggs.

At Fat Rooster, we use commercially made banks of nest boxes that are easily moved when the coop is being cleaned. The boxes are lined with absorbent pine shavings and then stuffed with hay to simulate a wild bird's nest. Some coops are designed so that nest boxes can be accessed from the outside for egg collection. This is ideal in smaller-sized coops, although flighty breeds may decide to exit the nesting box out the wrong side when their eggs are being gathered.

Some poultry keepers prefer to cover the entrance to the boxes with cloth or a strip of burlap to create a darker environment in which to lay. This practice can also cut down on the birds' tendencies to begin eating their eggs should they become bored with confinement in inclement weather.

BELOW: A Black Star cross hen sits on eggs in a nesting box at Luna Bleu Farm. The farm has 100 hens laying eggs; farmer Tim Sanford figures they need to get five dozen organic eggs a day to make a profit.

Roosts

All coops for chickens that will be kept into adulthood (and not slaughtered as meat birds) need roosts. It is a chicken's natural instinct to fly up off the ground away from predators and sleep for the night. Given that they will spend up to 13 or 14 hours a day on the roost, you should anticipate a great deal of manure piling up underneath. Locate the roosts away from entryways and exits into the chicken yard, away from the feeding stations and nesting boxes. The most practical roosts are capable of being removed from the coop during cleaning. A mesh wire floor that allows droppings to fall into a pan underneath the roost makes the manure easily transported outside on a regular basis. Another option is to have the mesh wire floor lead directly to the outside for manure removal.

The roosts themselves can be as simple as a branch suspended above the coop's floor. A more practical roost is one constructed with 4-foot lengths of lumber (such as 2 x 4s) bolted together

BELOW: Black Star cross hens roost in an unused greenhouse at Luna Bleu Farm. While concentrated manure is composted, the chickens' presence helps plants that are ready for harvest at least four months later.

at a 45-degree angle. Parallel pieces of lumber are then run perpendicular to the 2 x 4s, spaced about 24 inches apart to form an A-frame shape.

Litter

The lining, or litter, you choose to use will play an important role in the ease of cleaning out the coop later on. While hay or straw seems like the most natural choice, it can hold moisture, become stale and moldy, and be very difficult to remove once it has become soiled. Cornhusks were historically used, but are not readily available as bedding any longer (they are now commonly ground for animal feed).

Shredded newspaper is becoming more available, but it must be changed frequently in order to keep the interior of the coop dry.

RIGHT: Cornish Rock cross meat birds begin to wander from one of two portable coops early in the morning at Back Beyond Farm. While their manure left behind on the field improves the grass, the coop does need to be moved before too much manure accumulates and burns the pasture.

Some birds are also attracted to the newsprint and may consume it rather than their feed.

Dry softwood or pine shavings sold in bales at most feed stores are the ideal litter choice for chickens, though they are also the most expensive. The shavings do an excellent job of neutralizing the ammonia within the droppings and drying out the manure so that it is more easily handled. Shavings can also be used directly from the sawmill, but keep in mind that this litter has most likely not been kiln-dried, and will not be as good at absorbing moisture. It may also contain hardwood shavings, which tend to darken with moisture, making a less attractive interior in your coop.

A Barred Plymouth Rock cross rooster roams amongst the flower beds at Fat Rooster Farm. Its final destination was the freezer because the farm had too many roosters.

(4)

Feeding Your Chickens

My sister and I would sometimes lie in our beds and strain toward the northeast window, the covers pulled tightly round our shoulders, to listen for the spring crack of the ice going out on the river.

My little sister, Laura, was there too, just three years old, in the far room that looked out on the dead elm, where she bounced on her bed and sang, "Wookapecka, wookapecka" the day the flicker alighted on the tree. The dim light would just be coming in now, crepuscular sweetness, stung still by cold.

I remember the waking. Our eyes opening at dawn. The pony, in his stall, stamping and nickering at the hay thrown down, clouds of emerald dust streaking through the first threads of light.

Then outside, we girls, stamping and stabbing at the ice in the pond to make a hole, while the ducks waddled impatiently toward us through the snow. We watched, as the chickens reluctantly unveiled their heads from their feathery blankets to mumble sleepy thanks for the cracked corn and the compost. Afterwards, we trudged back to the farmhouse, stamping our boots clean before coming in to breakfast.

Finally, the pale skies yielding cold yellow light and the rooster's first crow. Then, over our shoulders, before we closed the mudroom's door, the river out, clanging loose those ice chains of winter, rushing free.

The way you choose to feed your chickens will determine everything from the quality of your meat to the hatchability of the eggs you save to incubate. Your feeding regime should be tailored to reflect the breed and age of bird that you're raising,

BELOW: A Barred Plymoouth Rock pullet snacks on weeds pulled from the garden of Jennifer Hauck. Hauck and her partner, Alex Cherington, raised them from chicks. It is Hauck's first time with chickens. "I wanted a constant supply of fresh eggs," she said.

the environmental conditions that the bird is being reared in (i.e., confinement versus free-roaming), and the outcome of the product you wish to obtain (i.e., meat, eggs, or show-condition individuals).

Commercial poultry growers rely on growing their birds using the least amount of feed and time (maximum feed conversion efficiency) in their poultry houses to increase profitability. To do this, rations are often manipulated with low levels of antibiotics (called *coccidiostats*) to control the single-celled protozoans that cause the disease coccidiosis. Amprolium is the most common additive in feed, though Monensin sodium, Lasalucid, and Salinomycin are also used. Roxarsone, an arsenic compound, is sometimes used to speed growth; other additives can include mold preventives and preservatives that stabilize vitamins and minerals vulnerable to temperature and humidity fluctuations. In some cases, human-grade food by-products make up a portion of the ration.

While antibiotics, coccidiostats, and other additives may be necessary in large, crowded poultry houses where the birds have no access to pasture and clean water and food, there's something about green grass and sunshine for preventing disease and boosting the immune system that they haven't been able to put in a bag yet.

The USDA maintains that Roxarsone is made up of an organic arsenic compound that is less toxic than the cancer-causing form. However, the EPA standards suggest that a child eating 2 ounces or an adult eating 5 1/2 ounces of cooked chicken liver weekly from chickens fed this diet may be exposed to arsenic levels that could cause neurological problems.

Antibiotic residues in meat can cause human health problems like allergic reactions, gastrointestinal upsets, and even resistance to antibiotics used to cure illness.

If your birds are provided with clean pens and pastures, fresh water, and uncontaminated feeding stations, many of the problems that require the use of these additives can be avoided right up front. Many companies offer feeds without additives, and organic grain companies comply with strict standards that regulate what goes into their rations.

Nutrient Requirements

Chickens require water, protein, carbohydrates, vitamins, fats, and minerals in their diet. The easiest way to supply these necessary dietary components is to feed your chickens a commercially available ration, although you can mix up feed for your chickens yourself using bulk ingredients. Feed stores are incredibly knowledgeable with regard to the requirements of your chicks, be they pullets for laying, broilers for the kitchen, or breeders for show. There are both conventional and organic feeds available in many parts of the United States.

Water

Water is essential for digestion and as a means of transporting food throughout the chicken's body and eliminating waste products. Because chickens don't have sweat glands, they need liquids to regulate their body heat. Chief sources of liquid can be fresh water, liquid milk, and fresh vegetation.

Carbohydrates and Fats

Carbohydrates are necessary in a chicken's diet for the production of fat, heat, and energy. In turn, fat is called upon, especially in times of stress, for heat and energy. Cereal grains and their by-products are the most common sources of carbohydrates in poultry diets.

Proteins

Milk, meat scraps, fish meal, soybean meal, cottonseed meal, and corn gluten meal all provide protein which is necessary for growth and repair of body tissues, the formation of eggs, and the production of fat, heat, and energy.

Vitamins

Necessary for health, growth, reproduction, and the prevention of nutritional disease, vitamins are found in a variety of sources. Green grass and other forages as

LEFT : Barred Silver meat birds eat a grower mash in the pasture at Fat Rooster Farm. The author chose the birds because they are easier to raise in a pasture than other varieties.

RIGHT: Black Star cross hens drink water from a tub at Luna Bleu Farm. The farm uses different colored chickens to be able to tell their age; later in the year, they were raising white chickens.

LEFT: With a bar across the box to prevent them from sitting in the feed, a pair of chickens eat pelleted layer ration in the coop.

RIGHT: A Rhode Island Red pullet pecks at zucchini from the garden in the enclosed pen outside a coop owned by Cloë Milek and Karl Hanson on their 2/3-acre lot. The couple has been raising 20 hens for eggs for the past seven years—they have avoided keeping a crowing rooster to keep the peace with their neighbors.

well as whole grains, wheat, corn, and milk contain an abundance of vitamin A, B, E, and K; sunlight and fish-based oils provide vitamin D.

Commercial Rations

Feed rations come in mash, crumbles, pellets, and scratch form. Mashes are finely ground so that they are easily picked up and digested, especially by young birds and chicks. Crumbles are intermediate in size when compared to mash and pellets. They tend to be less dusty than mash, and easier to pick up and swallow than pellets. Pellets have been partially heated within a pelletizer so that the grain particles are stuck to-gether and formed into ¼ to ½ inch-long tubes. Depending on the age of your birds and other environmental factors, one may be better suited for your situation. Each poultry keeper will have his or her preference; some like pelleted feed because the chickens have a harder time picking out the pieces of corn and filling up on the "ice cream" part of the diet. Others prefer mash or crumbles, because these types of feed take the chick-ens longer to eat and keep them busier and less likely to become bored and pick on each other. I like supplementing scratch feed with my maintenance ration for my grown layers during the months when there is bountiful forage, light is abundant, and the chick-ens don't need to use a lot of energy to keep warm.

Commercially prepared rations are specially formulated for new chicks, meat birds, pullets, and established layers or breeders. Each feed company has specific recom-mendations for how long to feed each type of ration, depending on your desired outcome, but you can use the following table as a general index. Above all, don't attempt to feed rations formulated for chickens to young game birds, guinea hens, turkeys, or waterfowl—the rations are usually lower in protein than these birds require, and in some cases, the ingredients can actually be toxic.

Name of Ration	Chicken Type	Number of Weeks to Feed
Chick Starter (20 to 24% protein)	Broiler	0 to 3 weeks (hybrids)
		0 to 6 weeks (heritage/range breeds)
	Pullets	0 to 6 weeks (commercial breeds)
		0 to 8 weeks (heritage breeds)
Poultry Grower (16 to 20 % protein)	Broiler	3 weeks to butcher (hybrids)
		6 weeks to butcher (heritage/range)
Poultry Grower (14 to 16 % protein)	Pullets	6 to 20 weeks (commercial)
		8 to 22 weeks (heritage)
Layer Ration (16 to 18% protein)	Mature Chickens	20 to 22 weeks

Making Your Own Ration

The advent of bagged grain for chicken feed is a relatively new one. As recently as the 1930s, chickens were fed the chaff leftover from the wheat harvest, the soured milk, the rotted vegetables stored in the earth cellar, and the excess produce from the farm's garden. Prior to the 1970s, poultry books were filled with elaborate recipes for feeding chickens. Everything from beet pulp to *tankage* (a fancy word for meat scraps and meat by-products) made up the rations. As poultry breeds became more specialized for egg and meat production, however, their dietary needs also became more specialized. Now, feeding only cracked corn to a Cornish Rock cross that is confined to a litter-lined coop would be a death knell. Their bodies require far more protein than the carbohydrate-rich corn can supply, and they no longer can supplement needed vitamins and minerals by foraging for insects, mice, and vegetation on pasture.

The practice of feeding meat scraps to chickens also became frowned upon, perhaps due to their rising costs, or concerns for rancidity and sanitation, and soybean meal is now the preferred protein source in a chicken's diet. Chickens, however, are omnivores. When left to forage, they will seek out living creatures to eat before they touch a trough of grain (one of my favorite pastimes is watching the meat bird chicks play tag with each other when they've found a moth or earthworm).

If you choose to make your own ration, you need to make sure that the proper energy requirements for your birds are being met. Scratch feed, a mixture of corn, barley, and wheat or oats, can provide carbohydrates in abundance, though it is severely lacking in protein. All too often, beginner poultry keepers will offer just cracked corn to their flock and become perplexed when the chickens stop laying eggs or die because of malnourishment.

One simple ration might be per 100-pound bag the following:

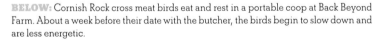

BELOW: Cornish Rock cross meat birds eat and rest in a portable coop at Back Beyond Farm. About a week before their date with the butcher, the birds begin to slow down and are less energetic.

65 pounds coarsely ground corn, oats, or wheat
7 pounds alfalfa meal or finely chopped hay
5 pounds meat scraps, fish meal, or soybean meal
13 pounds dried peas or additional soybean meal
6 pounds crushed oyster shell or limestone
3 pounds bone meal
1 pound trace mineral salt
½ pint cod-liver oil

Another alternative to making your own ration is to start with a lower-protein commercial diet and supplement it with other sources of protein (like milk, meat scraps, vegetables from your garden, or any other protein and vitamin sources you may have in excess).

As a general rule, any ration without preservatives in it should not be kept around for longer than three months, especially during warmer, humid months. Many commercial feeds contain mold inhibitors, but organic feeds do not. High heat, humidity, and old rations are a recipe for disaster if fed to your chickens.

Supplements

Depending on the breed of chicken and how they are maintained, your flock may need supplements to their feed. Typically, supplements are offered free-choice (left available in a separate container for chickens to eat at their own will) rather than added directly to the prepared ration. They can be left in the chicken coop in a rubber tub, or supplied to birds on pasture in a covered range feeder. I normally worry about supplementing only during the inclement weather months of winter, when the birds are confined and unable to do their own supplementing out on pasture.

GRIT. All chickens need grit to crush the food that they eat. Naturally, chickens swallow small stones and gravel as they forage outside, which become lodged in an organ called the *gizzard*. The gizzard is a well-muscled, kidney-bean-shaped organ that lies just in front of the stomach and acts to pulverize whole grains, beetles, and other hard food that the chicken eats. If confined to areas where grit is not obtained naturally, the chickens will need to be

ABOVE: A Golden Comet hen lays in a tub of a mineral mix supplement Ray Williams had put out for the chickens at Back Beyond Farm. Williams had read the supplement would ease the demands on expensive grain, but had not found it to work very well. He said sour milk was much more popular.

supplied with a commercially available mix that is available to them free-choice.

CALCIUM. If your hens' eggs are thin-shelled, you may need to supplement them with calcium. Ground limestone or oystershell is available commercially and can be fed in a separate tub. Don't feed eggshells to chickens unless they are finely pulverized and don't resemble eggs; chickens can pick up nasty egg-eating habits if they develop a fondness for eating eggshells.

KELP MEAL. Relatively new on the livestock feeding regime, kelp meal is rich in vitamins, minerals, and essential amino acids. I feed my laying hens free-choice kelp meal during the winter months, when green forage and insects are not readily available.

VITAMIN POWDERS. There are many commercially available supplements that can be added directly to the drinking water. These are especially beneficial for chicks being shipped by mail during the first few days of life in their new surroundings. They're also useful during high heat and humid conditions, when antioxidants might become rancid or deficient in the ration. Follow the directions carefully; vitamin A can be toxic to chickens in excess. Also, many vitamin and mineral supplements are not approved for use in certified organic meat and egg production, so consult the list of approved products before using any if you are producing organic meat and eggs.

How to Feed Your Chickens

Proper feed storage will reduce the chance of spoilage, as well as discourage an abundance of skunks, rodents, birds, and insects that are attracted to the free lunch.

Again, mold is detrimental to chickens, and moisture and humidity will break down the nutrients in the grain, leaving it unpalatable and possibly toxic to your chickens. A plastic or galvanized trash can will store grain nicely; open bags can either be emptied into the cans or left open inside (be sure not to leave the little strings that come off the bags lying around; chickens will invariably find them and wrap them around their feet).

Commercially manufactured feeders such as troughs or canisters are designed to hang from the coop's ceiling to feed your flock. They hover just above the ground so that rodents have a harder time getting into them, and chickens are discouraged from naturally scattering the grain around with their feet. Certainly, chickens can be encouraged to forage on pasture by scattering the grain amongst the vegetation, but it will also increase the amount of waste.

Chickens like routines. They are easily put off when things change, like weather, length of daylight, or the time of day that they're fed. It's best to feed your chickens within an hour or two of the same time each day; they'll lay eggs more consistently if they're laying hens and put on weight more evenly if they're intended for meat. I like to feed the birds in the early morning, after daylight has begun, and again at least two hours before sunset, so the birds have enough time to forage and then return to their roost, fed and happy for the evening.

If you are attempting to reduce the amount of prepared food they consume to take advantage of forage or garden surplus, delay feeding their grain ration until midday (never withhold water from them at anytime, even prior to slaughter). If they're confined to the coop and a small chicken yard, it might be better to feed them free-choice grain and supplement treats from the garden instead. I supply young stock from the time they are chicks with a summer squash or two from the garden to encourage them to eat more than just grain (they may not eat it right away, but after a few days, they usually associate the treats with good taste). Keep in mind that commercially prepared rations will put the most weight on your chickens the fastest and allow your laying pullets to mature the safest. You might save on grain bills by providing your chickens with other foodstuffs, but it will take longer to reach the end product. At Fat Rooster Farm, the trade-off is an acceptable one, as the goal is to decrease the inputs from off-farm while producing higher-quality, pasture-raised meat and eggs.

Laying chicks will require a different feeding regime than chicks that are destined for meat. Remember that a laying hen should develop slowly so that her body will be able to

ABOVE: A variety of Plymouth Rock cross and Barred Silver meat birds gather around the waterer in the pasture.

withstand the stresses associated with laying eggs for the rest of her life. A chick destined for consumption will be pushed to grow as quickly and efficiently as possible to achieve the best conversion of feed to meat production possible.

If you decide to raise the two together, you'll need to slaughter the meat birds first and then change the laying chicks' ration to one with less protein. Laying chicks should not be given food that is too high in protein because it will force them to mature too quickly. A hen forced into production at too early an age will consistently produce smaller eggs and be prone to prolapse (see Chapter Ten).

A simple way to slow the growth of laying-hen chicks is to introduce a whole grain like oats or wheat into their ration after they reach 10 weeks of age, about the time the meat chicks are slaughtered, and feed it combined with their grower ration until approximately 20 weeks of age. Then slowly change them over to layer feed (which contains higher calcium) for maintenance.

I prefer to raise my laying hens separate from my meat birds. The laying hens seem more precocious and tend to harass the slower, more docile meat birds, and it's easier to keep the feeding regime straight by having them separated. If you've decided on a dual-purpose breed (a non-hybrid that can be raised for both meat and egg production), it's easy enough to slaughter the roosters at 10 to 12 weeks of age, then add a carbohydrate source such as oats to the remaining hens' ration to slow their growth and maturity.

How Much to Feed Your Chickens

The amount of feed that your chickens will eat depends on several factors:

- The breed—a Leghorn laying hen will eat less than a Jersey Giant.
- The age—a 12-week-old Cornish Rock cockerel is an eating machine, whereas a 12-week-old Buff Orpington cockerel will not eat nearly as much.

BELOW: Lucy, a three-month-old Buff Orpington/Barred Rock cross pullet, forages in the backyard pen at the home of Geoff Hansen and Nicola Smith. The couple decided to try raising chickens for fresh eggs and to give their seven-year-old daughter a taste of farming.

- The season—chickens will require more energy to keep warm in colder months and will eat more. They'll also require more processed rations when they're unable to supplement their diet with fresh greens and insects. Chickens are usually more tolerant of the cold than heat, and their appetites suffer less when it's cold.

- The management technique—if your chickens don't have access to free range or pasture, they'll need to rely on the food you provide them.

- Chickens that are molting or at the bottom of the pecking order tend to eat less. Chapter Five explains the pecking order in more detail; basically the pecking order maintains the flock's social structure, where the most dominant bird, sometimes a hen, sometimes a rooster, rules the roost. All other birds within the flock fall into step under this bird in order of dominance to the other members of the flock.

I like to offer my chickens free-choice rations, meaning they are available to them at all times. Along with their commercially prepared ration, they are supplemented with treats—tomatoes and sweet corn gleaned from the fields, leftover salad greens, the meat scraps from last night's dinner. The treats become especially important during the winter, when their outdoor activities become restricted, and they're more likely to become bored. Sometimes I'll freeze the vegetables so that it takes the chickens longer to eat, keeping them occupied for a longer period of time. Cracked or whole corn can also be used as a treat, but it is nutritionally deficient, like scratch feed, and should not be used as the main ration.

Don't feed your chickens your eggshells unless they have been crushed beyond recognition. As noted earlier, feeding your chickens eggs is a good way to start a bad habit of having them eat their own eggs.

You can keep your chickens on a more restricted diet. Chickens intended for show are fed only as much feed as they can consume in 15 to 20 minutes. This trains them to associate human contact with the reward of food, making them more docile during judging.

ABOVE: A stray egg was left in the pasture by a hen at Back Beyond Farm. An errant egg comes about because the farm's portable coop is not closed at night, but it is protected from predators with an electric netted fence.

I make sure to feed chickens in several different locations so that even the lowest one on the totem pole will get enough to eat. If you find that the feed is disappearing as soon as it's put into the feeders, you're probably not feeding them enough or you may have a rodent problem.

A Black Star cross hen naps on a bale of hay at Luna Bleu Farm. Farmers Tim Sanford and Suzanne Long got this variety for their egg production. The color of the Black Stars also helps them tell their age difference from the younger chickens, who are white Leghorns.

" I find most henkeepers fall into two categories: Those like me, who have a few fowl and are obsessed by their personalities and the flock dynamic. The others, charmed by every breed, buy them all and leave them to get on with life, ending up with far too many birds, broods, and crossbreeds. This strategy is completely free range with little human interference, and even fewer eggs for the kitchen. It is glorious fun for a while but eventually you end up with mostly cockerels plus a few beleaguered hens who are abused and die off, leaving a posse of badly behaved louts. "

–FRANCINE RAYMOND, *THE BIG BOOK OF GARDEN HENS,* **2001.**

⑤

Keeping Chickens for Eggs

Keeping chickens for eggs is one of my earliest childhood memories. Chantecler the rooster and his two hens Penelope and Hortense, three regal Rhode Island Reds, strutted proudly around our chicken yard. Then came Bright Eyes, Fern, Lily, Funny Face, and countless other big red hens. Scattered among them were Wyandottes, Brahmas, Golden Comets, Black Sex-Links, Mille-Fleurs, Ameraucanas, and Australorps. As kids, my sisters and I would wait eagerly for the hens to lay their eggs and quickly snatch them up, scraping our names into the wet bloom. After they had dried, the names would magically stand out on the shell, like an Easter egg that has been waxed with a message before coloring.

We kept laying hens to provide us with eggs, but I really think my parents kept them as a way to teach us responsibility and self-sustenance. Laying hens are easily integrated into any small family operation that also has a small garden and cans or freezes zucchini and tomatoes to feed themselves in the winter. Chickens are easy enough to keep without becoming overwhelming, and anyone who has volunteered to feed the cats while you're away on vacation is usually willing to throw grain to a few chickens in the pen, especially if fresh eggs are part of the deal (they carry a lot more bargaining power than zucchini).

Keeping Laying Hens

Laying hens in a larger enterprise employ similar husbandry practices to those used by the backyard poultry keeper. Differences between small- and large-scale flocks might be in the breed of hen chosen, the feed they are given, and their living quarters.

WHICH LAYING HEN BREED SHOULD I CHOOSE? Each breed of chicken has a different rate of lay based on body type, ability to perform under certain environmental conditions, ability to convert feed efficiently to egg production, and other factors. The Chicken Breed Chart outlines many of these characteristics. Generally speaking, commercially developed hybrids produce more eggs. Most of these have been developed using the Leghorn, a breed in the Mediterranean class of chickens. They are flighty, small-framed birds that convert

LEFT: Whitey, a White Plymouth Rock, wanders amongst the bushes at Carol Steingress's and Rick Schluntz's home. The couple have been keeping 5 to 6 laying hens in their small in-town yard for the past 10 years. "I think they're so interesting and curious," Steingress said. "We're hooked."

feed more efficiently to eggs than other breeds. Leghorns are not the best to choose if a dual-purpose bird is what you're after. The carcass of a Leghorn can't compare to that of a Rhode Island Red or a Plymouth Rock when it comes time to cull them from the laying flock due to poor laying performance.

Egg color may also factor into your decision. It took us years for our customers to accept our white eggs. To this day, people still ask us if they are nutritionally equal to brown eggs. Now, we mix our cartons, so that each one contains white, brown, and green eggs.

Generally, the color of a chicken's earlobe indicates the color of egg it will lay; white earlobed hens lay white eggs, white red earlobes indicate a brown egg layer. This rule runs into trouble when you have Silkies, whose earlobes are a robin's-egg blue.

How Eggs Are Laid

When a female chick is hatched, she is equipped with two ovaries. The development of the right ovary stops in order to accommodate space for development of her eggs. Hens are hatched with the capacity to produce more than 4,000 eggs in a lifetime, but even the hardiest of hens, like the Leghorn, will lay just under 300 eggs in her first year of life, and then just over 200 eggs the year after. A pullet's first eggs will start out small, but by 30 weeks of age, they will reach their normal size. As a hen ages, her eggs will grow larger. Hens can lay eggs for more than a decade, but most are culled from commercial laying flocks after 11 to 24 months, when production decreases substantially.

Chicken Scratch

Are white eggs as nutritious as brown eggs?

The poultry demographic is clearly segregated. In New England, there are more brown eggs in the supermarkets, while in the Grain Belt and the South, white egg layers are more popular. The New England adage, "Brown eggs are local eggs, and local eggs are fresh" was true when the white eggs bought at the store were shipped from farther away. With modern-day transportation, this no longer applies, and on your farmstead all eggs, be they brown, white, or green, will share the same quality. There is no difference in the cholesterol content in brown versus white eggs, but free-range chickens (i.e., pastured, and allowed to forage at will) will produce eggs higher in omega-3 fatty acids than confined breeds. Confined birds can, however, be given feed supplements to increase omega-3s in their eggs.

LEFT: A Golden Comet hen sits in one of the nesting boxes of a portable coop at Back Beyond Farm. The coop has five nesting boxes – farmer Ray Williams said there should be one nesting box for every five laying hens.

Inside the hen's body, attached midway along her backbone, is the ovary and its cluster of undeveloped ova (the yolks). When the hen reaches maturity, each yolk comes to full size, and the follicle encapsulating it ruptures, expelling it from the ovary into the funnel of the oviduct. The yolk travels to the infundibulum, where fertilization takes place if a rooster has run with her. The chalazae, two twisted white cords that hold the yolk centered in the egg, are then formed, and the majority of the albumen (egg white) is added as the yolk twists its way through the passageway to the magnum. In the isthmus, the shell membranes are formed. In the uterus (or shell gland), the shell itself is formed, a process that takes about 20 hours to complete. Afterwards, the egg moves into the vagina, where it rotates from small end to large end facing out. A lubricating fluid, called the cuticle or bloom, coats the egg and eases it out of the vent. This quickly dries and seals bacteria and dirt out of the egg.

The hen will lay her egg one hour later each day until dusk is reached. Then she'll wait until morning to restart her cycle again. In their first year, most hens will not stop laying unless they become broody, they begin to molt, or the number of daylight hours decreases to less than 13 hours per day.

Raising Pullets

Believe it or not, rushing laying hens to maturity by feeding them a commercial ration designed to grow birds quickly like that fed to meat chickens can result in birds that lay smaller eggs throughout their life, have less vigor, and may molt their feathers too soon and cease laying prematurely. Pullets that are pushed to lay too quickly may also suffer from egg binding or prolapse. Eggs that are too large for the pullet's reproductive tract can get lodged (egg binding), which, unless removed, can result in the death of the bird. A large egg can also force the tender, pink skin just inside the vent to balloon out (prolapse). Since chickens are attracted to the color red, they will pick at the site, causing it to bleed and become infected. If the bird is not separated from the rest of the flock, it will be cannibalized by the other birds attracted to the blood and open wound and will very likely die from infection.

Chicken Scratch

Omega-3s

We sell our eggs at a farmer's market in an affluent town. One customer approached us, well dressed and well educated, inquiring about our eggs. Our chickens run, pell-mell, wreaking havoc in my sparse perennial beds, in the neighbor's yard, dodging the school bus, gobbling up fresh greens, insects, arthropods, even baby mice if they find them (chickens are truly omnivores). She asked us what we give our chickens to increase the omega-3 fatty acids in the eggs. I said that we gave them nothing, that they obtained these nutrients naturally, foraging on pasture. Nonplussed, she left, in search of artificially supplemented omega-3 fatty acid–infused eggs. In fact, hens raised on pasture have higher levels of omega-3s as well as vitamins E, A, and beta-carotene than hens raised in confinement.

It is best to raise laying pullets separate from cockerels or pullets intended for meat. The stress of constant harassment from cockerels or competition for food from the faster-maturing meat birds can delay the laying pullet's ability to reach maturity.

RIGHT: A Barred Plymouth Rock pullet warms herself in the sun in the doorway of the "Mini Coop" designed by Alex Cherington. The coop, with insulated floor space of about 12 square feet, is intentionally small to keep a small flock warm in the winter.

Access to pasture with fresh greens and sunlight will enhance a pullet's vigor. As the pullet reaches maturity, her rate of growth will decrease, and protein from a commercial ration should be decreased while the carbohydrate portion should increase to allow her to build a reserve fat layer.

Pullets will begin laying sometime between 18 and 24 weeks of age, depending on the breed. They will lay small, often blood-stained eggs at first, perhaps once every three or four days. By their second month of production, they should lay two normal sized eggs every three days. Pullets are best raised at a time of at least 8 to 10 hours of daylight, so chicks hatched in March and April will fare the best. As light diminishes, egg production will decrease unless a total of 13 to 14 hours of light per day is furnished. Pullets and hens will also lay best at temperatures between 45 and 80 degrees Fahrenheit.

Problems with Eggs

Abnormal Shells

SOFT OR NO SHELL. Indicates a problem with the shell gland. Stress can cause the egg to be laid before the shell has been formed; lack of vitamin D and lack of calcium can also cause the shell to improperly form. This problem is often seen in hens whose calcium needs are greater than in younger birds.

WRINKLED. Can indicate rough handling of hens, causing a second yolk to be prematurely released and bump up against the first egg forming. Wrinkled shells can also be an indicator of a respiratory infection.

BUMPY OR EXTREMELY CHALKY. Improper shell formation, sometimes seen in hens that have just

LEFT: A Cuckoo Marans hen, left, and Rhode Island Red hen roost in front of a nesting box at Fat Rooster Farm. The farm has 15 varieties of chickens, with 50 laying hens.

begun laying or in old hens, is also an indication of excess vitamin D. Don't use these eggs in the incubator—their uneven porosity won't allow the developing embryo to hatch (they're perfectly fine to eat, though).

Abnormal Egg Yolks

ORANGE OR DARK YELLOW—Most likely the result of the hen's diet. Leafy greens, carrots, marigolds, calendula flowers, whole corn, etc., can all affect yolk color.

BLACKISH-GREEN, OLIVE GREEN, OR REDDISH—Most likely diet-related. Green grass, acorns, and other green forages can cause these color changes.

PALE YELLOW—Lack of free-range forages and/or corn in feed.

BLOOD OR "LIVER" SPOTS IN YOLK—These are often confused with a developing embryo. In reality, an egg that has been partially incubated will look more like it is crisscrossed with a network of thin red lines. Blood spots or liver spots occur when blood or a small piece of tissue is released in the hen's reproductive tract before the shell has been formed. It can be an indication of vitamin A deficiency or a genetic defect that is inherited. While the eggs are perfectly safe to eat, they are unappealing. If a vitamin deficiency is not the problem, you may want to cull the culprit rather than risk putting off potential egg customers (or save the eggs for your own consumption).

DOUBLE YOLKS—Heavy-breed hens and older hens sometimes lay eggs containing two yolks. Rather than progressing normally through the ovary, the yolk combines with another yolk, and a shell is formed around both.

NO YOLKS—This is most common in pullets that have just made their first laying attempt. The egg contains no yolk and sometimes a speck of brownish or grayish tissue that the bird's reproductive glands have been tricked into treating as a completed egg.

Caring for Laying Hens

When a pullet reaches the end of her first year of life, she is considered a hen for exhibition purposes, but many people refer to female chickens as hens if they regularly lay eggs even if they are less than a year old.

The care of hens is similar to that of pullets, except that their bodies have stopped developing, and their vitality and vigor need to be maintained so that they continue to lay eggs. Laying hens on pasture will eat far less than those kept in confinement, and hens will eat less grain in the summer than the winter, when they need to keep themselves warm.

By the time hens have reached maturity, they will have worked out a distinct pecking order. Throughout their younger days together, this social organization governs how roosters relate to each other, how hens relate to each other, and how roosters and hens interact with one another. Maintaining a pecking order

BELOW: A Black Sex-Link hen roams the pasture at Back Beyond Farm. To add visual variety to their flock of Golden Comets, Ray Williams and Liz York added the breed. While they lay a lot of eggs, Williams feels Sex-Links can be protective while trying to remove eggs from the nesting box.

reduces stress within the flock. It is important to observe these social hierarchies to see that all of your chickens have access to drinking water, food, and a good place to roost away from danger. Setting up different feeding stations in the coop will allow even the most timid birds the opportunity for adequate food and water.

How to Determine Who Is Laying Eggs

A well-managed laying flock must include the removal of birds from the flock that are unhealthy or inadequate producers. Eliminating or *culling* these birds will reduce the cost of feed, improve your flock's production, allow more feed, water, and space for productive birds, and reduce the chance of disease spreading to healthy chickens. Observation and hands-on inspection can identify a good laying bird and assist you in choosing which hens to cull from the flock.

First, hens that are actively laying eggs should look like they've been working, not spending their days at the chicken salon. Their feathers may be broken or worn, because their energy has been put into feeding themselves and producing eggs, not preening sleek, shiny feathers all day. Their wattles and combs should be large, plump, bright, and waxy, not purplish or gray, wizened, and shrunken (Silkies are the color exception, with mulberry-colored combs as the norm). They should be active and alert, constantly scratching and picking for food, with sparkling eyes, often showing a very talkative demeanor. When you enter the henhouse, their eyes should be on you, wondering what treat you have brought them. Hens that are huddled, motionless, in a corner should be suspect.

Aside from observation at a distance, it is helpful to capture your hens and inspect them closely. There are numerous ways to catch chickens. If the hens are tame, simply scoop them up, keeping their wings closely pinned to their body to avoid harm to them and to your face.

A simple tool that is useful in catching timid hens is a *catching hook*, made up of a piece of sturdy wire that is hooked at the end. These are available from poultry supply outlets, or you can make one from Number 3–gauge wire. A length of 5 to 6 feet is ideal, with a sharply bent hook at the end. The chicken is caught by hooking her by the leg and quickly scooping her up. After she has been caught, hold her for a few moments before examining her to calm her and get her used to handling.

Another handy tool is a fishing scoop net. As long as the mesh in the net is fine enough so that a chicken can't escape, these work well for scooping up feisty fowl.

Clipping the hen's wings will also render them more easily caught. The primary flight feathers on one wing can be cut so that the bird becomes off-balance when trying to fly. Use scissors to cut the first 10 feathers on the wing, starting from the outside of the wing and moving toward the body of the bird. Wing clipping will only last until the new feathers grow in during the next molt.

Trap nests can also be used to capture hens for inspection. These nests have specially designed doors that close after the hen has entered to lay her egg. Trap nests are also a useful way to track each bird's production. Peruse poultry supply catalogs for nesting boxes equipped with trap doors, or use Appendix Four to find trap nest plans you can construct at home.

LEFT: The vent of a hen who is laying eggs is moist; a molting chicken's is dry.

RIGHT: The vent of a molting hen is dry and is not laying eggs.

Three physiological characteristics will help you determine whether your hen is laying. The first is the appearance of her vent, where the egg exits her body. Tip the hen up so that her head is facing toward the ground, her wings held against your body and the arm you are holding her in. Her tail will naturally tip forward. Part the feathers near her vent with your free hand. The vent should be swollen, full, moist, and pink, not dry, puckered, or purplish.

Next, place your thumb and middle fingers of your free hand on either side of her vent. Feel for two sharp bones on either side—these are her pelvic bones (these are different from the keel bone, which lies midline along the breast). They should be flexible and wide on a hen that is laying. If they are firm and almost joined just below the vent, she is most likely not laying.

Finally, a laying hen with yellow skin will lose pigmentation from her body in an orderly fashion, called *bleaching*. Bleaching occurs when the carotenoid pigment called xanthophyll is diverted from the hen's body to her egg yolks. The more eggs she lays, the more this pigment is depleted, resulting in her body parts becoming

BELOW: Ready to be gathered, eggs rest in a nesting box at Luna Bleu Farm. The chickens' foraging in summertime pasture benefits in better-tasting eggs. "People are fighting over our eggs," said farmer Suzanne Long.

paler and paler. Her body will lose its color in order from her vent, to her eye ring, to her earlobe. Then will follow her beak, her feet, and her shanks. By the time her shanks show signs of bleaching, she will have laid upwards of 180 eggs. Bleaching will not be a good diagnostic tool to detect laying in hens whose skin is black (like Silkies) or white (like many of the Asian and Continental class breeds), or in hens who have not been fed a diet that contains corn or rich, leafy greens, where xanthophyll is found.

BREAKING UP BROODY HENS

In the 1940s, broody hens in commercial laying strains were far more common than they are now. G. T. Klein in *Starting Right with Poultry* noted that the strain of layers he began with as an extension poultryman at the Massachusetts State College in Amherst (now the University of Massachusetts–Amherst) went from 100% broody to just 3% broody 25 years later.

Heritage breed hens tend toward broodiness more than commercial strains of Leghorns, Black Sex-Links, or other laying

breeds that have been selected and managed to enhance their non-broody characteristics. If you don't wish for your hens to stay broody and set on eggs, they should be "broken up." Otherwise, a great deal of egg production will be lost while the hen thinks she is "incubating" her imaginary clutch of eggs (remember, hens do not lay eggs while they are broody).

There are two techniques that are usually effective in breaking up a broody hen. Both rely on the hen's instinct to give up on a lost clutch of eggs so that she will live another day to start fresh. The first is to remove her from her nest and put her in a different coop. The change in scenery is sometimes enough to shock her out of broodiness.

Another method is to place her in a broody coop. This is a hanging cage with a wire floor. Usually three or four days is all it takes to get her hen brain off any thought of setting on eggs. Food and water can be provided to her throughout her time in the broody coop.

The sooner you break up a broody hen, the more quickly she will return to laying eggs. If she is prone to broodiness, you may want to consider her culling her, unless you're interested in hatching your own chicks.

FORCING THE ANNUAL MOLT

Hens will cease laying when they begin molting their feathers. When day length begins to shorten, it will trigger the hens to molt their worn plumage and replace it with sturdy feathers in preparation for winter migration

LEFT: A molting White Leghorn hen; While it is in the process of molting old feathers to grow new ones, a chicken's egg production slows down. The comb and wattles also become dull and shrunken.

RIGHT: A White Leghorn in "full lay." The comb and wattles of a laying bird will be bright and plump.

(even though chickens have long since stopped migrating like their wild cousins, their bodies are still programmed to molt with winter's onset). The best laying hens will molt later in the season, taking nearly four months to replace all of their feathers. During molting, hens divert the energy used to produce eggs into feather production and stop laying for about two months; commercial layers will slow their production, but rarely cease egg-laying entirely.

Chickens will molt in a specific sequence, beginning with the head, to the neck, back, and breast, with the tail and wings being molted last. Some hens will appear almost naked, molting their feathers all at once.

Poultry keepers who find it more economical to keep their mature layers rather than sending them to slaughter in exchange for fresh pullets will sometimes force their birds into an early molt. This practice requires great skill, as it essentially creates enough stress on the bird to force it to lose its feathers. Great care must be taken so that the conditions the birds are kept in do not cause mortality and a complete cessation of laying.

Older poultry management manuals suggest feeding laying hens forced into molt grain only, and then, after a period of eight weeks, returning them to a layer ration. All-night lighting is also required after the hens have been returned to their regular ration.

Alternatively, hens can be confined in a coop that is well ventilated and restricted to just eight hours of light a day. During this time, only free-choice water and oats and ½ pound of scratch feed for every 12 hens should be given. In two or three weeks, layer ration should be reintroduced, and light should be increased gradually to 15 hours per day.

CULLING LAYING HENS

In small flocks, where birds will certainly take on individual personalities, culling can be hard to follow through with. The decision to cull birds is purely management based. If you have decided that your flock serves more purpose as companionship than as a means of providing eggs profitably, then culling would only be necessary to eliminate unhealthy and diseased birds from the otherwise healthy flock. Normally, hens kept for the purpose of producing eggs are culled at the peak of production if they're not laying as they should (30 to 40 weeks of age, depending on the breed) and again at the end of their first year of production.

Introducing New Birds to Your Flock

Like the new kids starting school three or four months into the year, the arrival of new chickens, be they hens or roosters, to the laying flock will create a fair amount of waves. The established pecking order will be challenged, and each bird in the flock must now establish a relationship with the newcomer. Choosing a time to introduce new birds that is least stressful on the flock may make the transition easier. If yours is a commercial enterprise, it's best to overturn the entire flock at one time. Started pullets from April can take the place of the old layers in the fall, when those birds are close to completing their first year of laying. I find it useful to switch breeds from year to year, so I can tell which are the new birds

BELOW: White Leghorn hens, in foreground, and a variety of other chickens roost in the coop. The fifty laying hens lay about two dozen eggs per day.

and which are the old. If you are just introducing one or two new birds to the flock, put them in the coop at night, and then carefully observe the new birds to be sure that they have access to food and water. In free-ranging operations, transitioning new birds into the flock tends to be smoother because of the extra space allowance, though clipping the wings of the new chickens may prevent chasing them out of trees, horse barns, or haylofts.

Marketing Eggs

The modern commercial poultry industry is highly specialized to reduce waste, labor, and overhead. Size, volume, and marketing potential dominate management decisions, and profitability depends on efficient production of large quantities of eggs on a few large farms. Capital investments are enormous in these poultry plants, leaving most small-farm managers out of the running for competition.

BELOW: A meat bird hen rescued from a date with a butcher examines the nesting boxes at Fat Rooster Farm.

On the other hand, a small-scale project, which targets niche markets, can be sustainable and add to a homestead or farm in a number of ways. First, the addition of poultry products to other agricultural crops offered could attract customers that may otherwise turn to other sources for eggs. Second, poultry manure is often overlooked as a benefit of raising chickens, and the addition of this nitrogen-rich fertilizer will enhance soils and benefit other crops. Lastly, creating value-added products from eggs can increase revenue.

TO SUCCEED WITH YOUR POULTRY OPERATION, YOU MUST:

IDENTIFY YOUR MARKET. Who will you sell to? Local food co-ops? Neighbors? At farmer's markets or at your own farmstand? Be sure to discuss details and delivery, expectations, and your price with potential customers.

FIND A NICHE THAT IS UNIQUE TO YOUR OPERATION. Our niche is heritage-breed chickens that lay a myriad of colored eggs. We carefully mix each dozen so that it contains at least one chocolate brown or olive green egg. Customers will choose our eggs over our neighbors' just for their appearance.

CREATE VALUE-ADDED PRODUCTS TO INCREASE YOUR PROFIT. Consider making quiche, pickled eggs (see Chapter Nine), or pre-packaged containers of egg salad for sale. Be sure to review your local rules and regulations, as many states require prepared food licenses and commercial kitchens to make these products.

KEEP YOUR COSTS LOW ENOUGH AND YOUR OPERATION EFFICIENT ENOUGH TO REALIZE A PROFIT. This last is key to a successful business venture. Specifically, tracking flock production, feed costs, feed conversion (the amount of feed used to produce a dozen eggs; a good number is 4 pounds of food per dozen eggs produced), and feed efficiency (the cost of feed spent to produce a dozen eggs) will help keep your operation in the black.

One useful statistic is the flock's average production. At the beginning of the month, tally the number of hens. Each hen represents a hen-day, or a potential egg laid. Keep track of how many hens have left the flock (sold or died). At the end of the month, subtract the total number of hen-days lost from the total number of hen-days. Divide the number of eggs laid in the month by the number of hen-days, and average flock production is obtained.

For example, in a flock of 40 hens, one hen is lost on day 6 of the month of March and one is lost on day 10, for a total of 46 hen-days lost (31 days in March, subtract 6 = 25 hen-days lost; 31 days in March, subtract 10 = 21 hen-days lost for a total of 46). Forty hens multiplied by 31 (the number of days in March) = 1,271 hen-days, minus 46 (the number of hen-days lost) = 1,225 hen-days divided by the total number of eggs for the month. A good average flock production will range from 15 to 20 dozen eggs per hen per year.

If production statistics sound too complicated, and you are keeping chickens purely for the fun of it, you can still take in some revenue from what you produce. The difference is that you won't deem the operation a failure if it's not profitable.

Perdue, a Dark Brahma pullet, basks in the fall sunlight at Brianne Riley and Matthew Taylor's home. When the couple was married in mid-October, they included their chickens in their vows.

One of two Buff Cochin pullets owned by Brianne Riley and Matthew Taylor of Shelburne, Vermont scratches amongst the leaves in the yard. Riley said customers to the antique shop she runs on the weekends like to visit with the chickens.

"Most of the poultry marketed in this country comes from farm flocks, and is essentially a by-product of egg production."

–WILLIAM LIPPINCOTT IN *POULTRY PRODUCTION*, 1935

⑥

Keeping Chickens for Meat

In the early part of the twentieth century, the poultry industry was dominated by egg production. Today, poultry meat is big business, and present-day marketing to the mass consumer has perpetuated a dizzying array of definitions and deceptive descriptions that may not necessarily have anything to do with how the final product has been raised. Now that organic agriculture is the fastest-growing sector in the U.S. agricultural industry, everyone is trying to get a piece of the pie. What started in the late 1960s and early 1970s as a grass-roots effort concerned with treading lightly on the soil and based on philosophical principles and moral obligations is now a free-for-all, and marketing takes the prize: the consumer's dollar. Bandied around are terms like organic, all natural, antibiotic-free, no growth hormones added, free-range, free-walking, free-running, free-roaming, and cage-free.

ABOVE: Cornish Rock cross meat birds stop for a drink of water at Back Beyond Farm. The farm raises and sells eight batches of 90 to 95 birds apiece to sell to customers who visit them on the farm and at farmer's market.

Historically, Americans used "free-range" to describe animals that were permitted to roam freely, without containment. Now, most consumers equate free-range chickens with methods similar to those used up to the 1950s, where birds were rotated in large outdoor yards, capable of supplementing their grain rations with vegetation, insects, and soil micro-nutrients. They had direct exposure to sunlight, fresh air, soil, and rain, which stimulated preening (the distribution of natural oils on the feathers) and overall general health of the birds.

In fact, free-range and similarly misleading terms are not legally defined and conjure up visions of animals allowed to live their lives instinctually, unencumbered until they are humanely slaughtered for consumption.

The majority of commercial poultry grown for meat in the United States is not housed in the battery cages that many laying hens are kept in. They are kept in poultry barns, measuring up to 500 feet long and 50 feet wide, with a capacity to house 20,000 birds. Water and food are provided abundantly, through overhead nipple, cup, or bell waterers and feeders that are placed off the ground to eliminate feed waste resulting from a chicken's natural tendencies to scratch at the soil to find food. Ventilation, temperature, and light are carefully monitored to achieve the most proficient feed conversion and produce birds ready for slaughter at 4 pounds dressed weight (the weight of the bird after it has been plucked and eviscerated) in just over a month.

These poultry houses can legally refer to their birds as being raised by any of the previously mentioned terms without clear definitions. In truth, unless the company states that their birds are raised on pasture, with consistent access to the outdoors, there is really not a lot of difference between any of the conventional meat chicken facilities. They all have Cornish Rock crossbreeds, packed in large barns where the ability to peck at the soil or green grass does not exist. They're given conventional grains, which likely contain genetically modified varieties of corn and/or soybeans, and the birds are slaughtered in assembly lines at three to ten weeks of age. As in cattle feedlots and hog barns, conversion of feed to meat is the bottom dollar, not the individual's quality of life.

LEFT: A Barred Silver cross cockerel meat bird rests in the pasture. Meat birds who forage on grass have a richer diet and produce more flavorful meat.

Our food's safety underwent serious scrutiny after the foot-and-mouth and mad cow disease (bovine encephalitis) scares at the turn of the twenty-first century, and the nation's consumers turned to methods of farming that claimed to be healthier, safer, more humane. Suddenly, the adjectives used to describe the small-scale farmsteads became the semantics of agribusiness, and large corporations capitalized on consumer transformation.

Previously, individual states had private certifying agents that supervised the small, diverse farms across the country devoted to organic agricultural production. The USDA sensed a need to regulate what could be called organic so the consumer could be honestly educated, and today, farms claiming organic status for their poultry products are governed by the USDA's certifying agency. Unfortunately, there have been several attempts by powerful lobbyists to change some of the organic regulations, so many of the same concerns remain.

Raising Meat Chickens by Type
CONVENTIONAL MEAT BREEDS

Historically, roosters from chickens that were raised for egg-laying were slaughtered for meat consumption on the homestead as a by-product of the hatch. The birds were slow to mature and

ABOVE: A Cornish Rock cross meat bird forages on pasture at Back Beyond Farm. The chickens graze until their are about six weeks old and are ready for butchering at seven to nine weeks old.

weighed no more than 4 or 5 pounds at slaughter. Today, most meat chicks sold by hatcheries are hybrids referred to as Cornish Rock, Cornish Roasters, or Cornish Game Hens. These are the same breeds that are sold to commercial poultry producers like Tyson and Perdue. They have been bred to develop plump breasts, meaty thighs, and have feathers that pluck cleanly, leaving only transparent pinfeathers. They require high-protein rations, often needing supplements in

Chicken Scratch

Just as some breeds of cows have been developed for milk and others for meat production, hybrid chicken breeds have been developed for eggs and meat production. In the United States, the preference is for yellow-skinned flesh that is plump in the breast and thigh, but a chicken's meat can range in color from pinkish (as in the Catalana breed) to black (as in the Silkie).

"A chicken in every pot, and a car in every garage."

—HERBERT HOOVER'S 1928 solution to end poverty in this country

their drinking water to prevent leg and heart problems, and they're generally poor foragers. They are docile and gregarious, but making long-term pets of them is almost certainly a disaster. They have been selected for rapid growth and specific weight gain that will not accommodate longevity or the ability to naturally reproduce. Ordering straight-run (birds whose gender has not been determined) chicks tends to be most economical and will provide a variety of both larger and smaller dressed birds.

Commercial meat chicks should be raised separately from other chicks such as commercial laying breeds or heritage breeds. These more precocious birds are more active and may even cannibalize the slower, docile meat bird chicks.

Typically, hybrid meat bird carcasses are classified by their size, and not necessarily dependent upon sex or age of the bird. The hybrids that you raise at home are identical to the chicken offered for sale at your local grocery store:

CORNISH GAME HEN OR CORNISH ROCK GAME HEN. These are Cornish cross hybrids, not really game birds like quail or pheasant, and not necessarily hens. They are tender and young, weigh between 1 or 2 pounds, and are typically five or six weeks of age.

Raising game hens for sale requires a premium asking price, as most weight gain usually begins just at the time of slaughter (the cost of raising the bird to six weeks of age is much greater than the cost of raising it to 10 to 12 weeks of age, when it will dress 6 to 8 pounds). On the other hand, rather than waste chicks that are exhibiting signs of leg problems or who are injured, I have culled and dressed young chicks at this weight. They are excellent on the grill (see Beer Can Chicken, page 194), and culling them before they are lost in production can increase overall profitability.

BROILER OR FRYER. These can be pullets or cockerels, and both heritage breeds and hybrids make good broilers and fryers (a commercial breed that I have often used is a cross called a Buff-Silver). They dress between 3 and $4\frac{1}{2}$ pounds and are typically less than four months of age. At this age and weight, the birds are still considered tender and young and can be used for a variety of purposes.

ROASTER. These are usually Cornish-cross hybrids. Their flesh is still tender and moist, with flexible skin and pliable breastbones

A Buff Silver meat bird warms itself in the morning light.

and a finishing weight of between 5 and 9 pounds. Heritage breeds can be grown to this size, but it takes them twice as long to achieve the same weight and they're usually better in a casserole or a soup.

CAPON. Before specialized hybrid broilers were developed, caponizing was necessary if cockerels that had been separated from hens intended for laying pullets were to finish with tender flesh. The purpose of caponizing is to force the birds into growing plump and juicy, rather than expending energy developing secondary sex characteristics like combs, spurs, and testicles. They are surgically castrated at two to three weeks if they're hybrids, and at five to six weeks if they are heritage breeds. While the procedure can be performed at home, and caponized cockerels can still be purchased (See Appendix Four), raising capons is not necessarily economical or practical for a homestead. In 25 weeks, a heritage-breed capon can eat almost 40 pounds of feed, whereas an 8-pound hybrid roaster will dress nearly its equal in 12 to 15 weeks.

BELOW: A Buff Silver meat bird escaped its date with the butcher and is part of the coop. "He'll be good for company as a stew," said the author.

Heritage or Dual-Purpose Breeds

Heritage breeds of poultry are perfectly suitable for smaller-scale meat production. They mature more slowly, dress smaller (taking more like 13 to 22 weeks to dress out at 4 to 5 pounds), and their carcasses will not necessarily be as butter-yellow in color or as overly plump in the breast and thigh as their giant, white, marshmallow cousins. The meat will be firmer in texture with a stronger chicken flavor. They are heartier than the commercial hybrids, and can be adapted more readily to foraging operations. These birds do not suffer mortality as the hybrids do from frail hearts or legs that cannot support added breast weight.

An added benefit to raising heritage breeds of poultry is the ability to choose a dual-purpose breed, where males are butchered for meat, and females are kept as laying hens. If you decide to raise heritage breeds, the sexually maturing males should be separated from the hens at about 10½ weeks of age to avoid harassment and injury to the hens.

Preparing Your Birds for Butchering

When I was about 10 years old, my dad decided that our family would raise its own chicken meat. We'd had laying hens for as long as I could remember, but we'd never eaten any of the chickens that we'd cared for. They were, in fact, considered pets that were kind enough to provide us with eggs.

Despite stern warnings, the white cockerels that Dad bought and designated as meat chickens ended up with names, like Cry Baby, Heart Attack, and Sylvester.

On killing day, my sisters and I sobbed when the chicken's heads were lopped off in the backyard, and Mom plucked them in the kitchen sink. The memory of headless chickens flopping on the lawn and the smell of damp feathers in the house put an early end to my chicken meat enterprises.

Living in Brazil several years later, I was again confronted with people who raised the meat that they ate. Native poultry would be chosen at open-air butcher markets and brought home to be slaughtered. Nothing was wasted—the blood was caught in a bowl to be used in the sausage, the

ABOVE: Butcher Ralph Persons of Hardwick, Vermont, talks with his wife, Cindy (not pictured), while gutting chickens in their portable slaughterhouse during a visit to Fat Rooster Farm. Persons estimates he will butcher 17,000 chickens in 2008, up from 14,000 in 2007. Persons has been a butcher for 27 years.

feet and neck were stewed with collards for a flavorful broth. Even still, it made me squeamish to be so intimate with my food.

After attempting vegetarianism as an alternative to confronting my dependence on meat, I met Bradford and Donna Kausen, a husband-and-wife team from Downeast Maine who fished, vegetable farmed, and grew their own meat. They taught me everything from how to skin a deer to how to prepare roadkill partridge. Most importantly, I learned to accept the fear I had in possessing the ability to deem something living a personal food item. And while I feel the same gratitude when I pluck a ripe tomato from its vine, I'm not confronted with trusting eyes and the sounds of an animal that is trying to make a connection with me.

Why do I continue to eat meat if I am aware that I am directly responsible for this animal's demise? Because I am a meat eater, and because I want to eat meat responsibly. Because without killing meat to eat, I've still displaced wildlife by building the dwelling that I live in, and I've taken the long grass that harbors turkeys, deer, and Savannah Sparrows to grow my corn, lettuce, and squash.

I want to be ever-aware of this power I have, and to be afraid of it and to respect it. I feel that if we are to eat meat, we should know how it has lived its life and be responsible for accepting how and why it died.

Two Buff Silver Cockerels poke their heads out of a crate while the butcher prepares to do his work. While some farms do their own butchering, the author hires someone to come do the work because of how well he does the job.

There are two options available for processing the meat that you raise. The birds can be shipped to a slaughter facility and processed there by experienced butchers, or they can be processed at your home, either by a mobile processor or by yourself.

In Vermont, poultry processing facilities are few and far between. Fortunately, we have a very reliable mobile processor who comes to the farm. The birds are killed, plucked, gutted, and cooled within a few hours, and then it is up to us to finish preparing them for the freezer. There is less stress involved when the butchering is done on the farm; the birds remain in a familiar location and are handled less. There are several rules and regulations concerning the sale of meat that has been processed in an uninspected facility, however, so if you intend to sell the meat, do your homework.

The easiest way to find someone to process your birds may be to inquire at the local feed store, at the extension agency, or at a fellow producer's farm. Again, knowing how the birds will be processed and ready for the freezer is something best decided before you are actually ready to slaughter your birds.

If you've decided to transport your birds to the butcher, you should keep the following in mind:

LEFT:
Apprentice Whitney Taylor of Wellsboro, Pennsylvania holds Poopie Poo, one of the meat birds destined for the butcher. Taylor had become attached to the chicken—it was left in the coop with the laying hens.

- Reducing the amount of stress on the birds will decrease the chance of tough, off-flavor meat. Catching the birds and placing them in their transport cages after dark will reduce stress and the temptation for cage mates to pick on one another. Remain calm, and talk to the birds as you would in the daytime, when you are feeding them. Believe it or not, if you pick your birds up by grabbing their feet and carry them head-hanging-down to their transport cages, the risk of injury to their wings or thighs will be reduced, and the birds will generally be calmer. Padding the cages with extra bedding will reduce the amount of manure on their feathers and feet, which in turn will keep the processing water and the finished carcass cleaner.

- Caged birds generate body heat. Provide ample ventilation and keep them out of direct sunlight if you are shipping them to be processed during daylight hours.

- Don't feed your birds 12 hours before slaughter, so their digestive tracts will be empty, and the amount of manure they make will be reduced. Make sure to provide them with plenty of fresh water prior to transporting to slaughter.

RIGHT:
Whitney, left, and the author gather three meat birds for the butcher at Fat Rooster Farm

- Getting the birds to the butcher will require containers that cannot be used later to transport the processed meat, so think ahead as to how you will get your meat home.

If you are processing the birds yourself or a mobile processor is traveling to your farm, here's a list of points to consider:

- Start getting ready the night before. If a mobile processor is coming to your farm, be sure to confirm the time and date. Prepare all needed water and electricity sources, live-bird holding pens, containers for chilling processed birds, waste containers for composting feathers, feet, heads, and innards, and freezers or refrigerators to hold the finished birds.
- As you go through the process of harvesting what you've grown, try to consider all of the factors that will make this process as dignified and cruelty-free as you can. The more you can keep your birds calm and comfortable during the slaughter process, the better the meat will taste. Stress can physiologically change the meat's composition. One way to reduce stress is to keep the live birds away from the slaughter site. Birds are capable of seeing red, and distress calls and smells of blood can add to a stressful atmosphere. Aside from producing a better-quality product, one of the reasons to raise your own meat is to do so with respect for these living creatures.
- For birds that will be retailed, make sure that you have communicated with your customers. Some of them may be unfamiliar with how to handle a whole bird. Giving them cutting instructions or supplying the birds in quarters may generate repeat customers.
- Don't underestimate the by-products of the slaughter: livers, gizzards, feet, hearts, and necks. These can be used in specialty markets, for pet food, or value-added products like pâté or soup stock.

Basic Butchering Steps for Processing Your Birds at Home

I strongly recommend apprenticing with a skilled butcher or a neighbor who has already mastered butchering if you have never tried it. There is nothing more intimidating than attempting to humanely and effectively slaughter your animals without knowing what you're doing, especially when blood and feathers are flying.

The first step in slaughtering is finding a way of securing the bird tightly during the butchering process. A *killing cone* is the most efficient and least stressful method of securing the bird; it will protect its wings and thighs from bruising and keep the head stationary so that the blood will drop to the ground without soiling the feathers. Bleeding the bird out as thoroughly as possible is important; it reduces the risk of spoilage and makes a nicer-looking carcass. You should also have a very sharp knife to use for killing the birds.

If you are wet-plucking your birds, the scalding water should be ready to go and positioned close to where the birds are being killed.

BELOW: As helper Bev O'Neill, left, cuts off the chickens' heads while they are in one of three killing cones, butcher Ralph Persons guts the birds after their feathers have been removed in the plucker. Persons and his wife, Cindy, travel across the state with his portable slaughterhouse, butchering animals and birds on the farm.

If you're slaughtering more than three or four birds, it's helpful to have one person doing the slaughtering and another doing the scalding and plucking. After scalding and plucking, there should be another pot of ice water where the meat cools immediately after slaughter. Once you've worked out where the birds will be killed, scalded, plucked, and cooled, you're ready to start.

Place the bird head-first in the cone by dropping the bird in while securing its feet. Grasp the head by your least-dominant hand (right hand for lefties, left hand for righties). Fingers should be holding the head on either cheekbone, just below the bird's eye; I use my thumb and ring finger to do this.

The second step is deciding what to use to actually kill the bird. Some experts advocate using a .22. I'm a pretty good shot, but a chicken's head is a pretty small target, especially if it's moving around. There is also the added worry of where the bullet goes after it exits the chicken's head. Snapping the bird's neck by placing its head underneath the handle of a broom or rake and pulling up firmly will also work. However, the bird will flap vigorously, potentially bruising the meat in the wings and breast. They still need to be cut across the neck to bleed out. Axes are also popular, but again, the bird will flap about, and blood spatter can't be contained well.

For this reason, I am more comfortable using a knife to sever the jugular veins and bleed out the bird. Taking a knife with a long, thin blade, such as a boning knife, insert the tip close to where you are grasping the bird with your thumb, push it all the way to the other side, and swiftly pull the blade out away from your fingers. You should see two prominent streams of blood spurting forth, where you have severed the jugular veins. If you don't, swipe deeper in the area that you are grasping. You should continue to hold the bird throughout this process until the wings grow limp. Some people prefer to cut the head cleanly off. I find it harder to control the bird from flapping without being able to grasp its head.

When a free flow of blood is observed, release the bird's head. At this point, it may shudder or flap in the cone. The bird is ready to continue processing after it has grown limp, and the head droops freely.

ABOVE: Bev O'Neill removes feathers missed by the plucker.

Pithing—shoving the point of the knife into the back of the brain through the roof of the mouth—is said to release the feathers for easier picking, and some say that it immediately stops any sensation of pain. If you choose to try pithing, hold the bird's head after it has been cut for bleeding out, and stick the point of the knife up into the groove in the roof of the bird's mouth, toward the back of the brain. Make a small turn, then pull the knife straight out. The bird should then be allowed to hang for bleeding out until it is limp (about five to ten minutes).

The bird is now ready for removing its feathers, called *picking* or *plucking*. You can *dry-pluck* the bird by suspending it and picking the feathers by hand. I have always found this to be a more tedious process than wet-plucking, but it works when hot water for scalding is not available or if you don't want to boil a bunch of water for just one or two chickens. Removing the feathers from the bird's body should be done in the same sequence as for birds that are wet-plucked, described below.

To *wet-pluck*, remove the bird from the cone and, grasping its feet, dip it in the pot of hot water (125 to 155 degrees Fahrenheit, depending on the age of the bird, with the lower temperature for younger, tender-skinned birds). Swish the bird back and forth and up and down, as you hold its feet, to encourage the water to penetrate the feathers. To test if it has scalded long enough, pull at the flight feathers on the wing; they should pull out freely. If they don't, soak the bird longer, and check the temperature to make sure it is hot enough. I usually scald the bird no more than 45 seconds to a minute.

When the feathers pluck freely, remove the bird from the water. Suspend it at a comfortable height or place it on a level surface. Begin plucking the feathers on the wings, then shoulders, back, neck, breast, legs, and finally thighs. This sequence will follow the amount of subsquent blood loss from the various regions of the bird and will guard against bruising from blood that has pooled due to improper draining. To pluck the feathers, use your thumb to lift them away from the body and pull backwards, opposite of the direction in which they naturally lie. Make sure to dunk the bird in a bucket of cold water during plucking if the task takes longer than five or ten minutes. If the skin overheats, it will cook and may tear or become wrinkled (if this happens, the meat won't be ruined; the finished bird just won't look as nice).

After plucking, thoroughly rinse the bird. Remove it to a flat surface and, using a sharp knife, remove the head and legs between the bone joints.

Flip the bird on its belly, neck facing you. Slit the skin up to where the shoulders meet. Pull the skin down, and expose both the gelatinous esophageal tube, and the firm, corrugated, ridged trachea. Pull these both away from the neck, toward the shoulders.

Now flip the bird onto its back. The crop, a sac which holds the undigested food until it makes its way to the stomach, lies on the right side of the bird's breast; you can trace back to it by holding the esophagus that you have pulled from the neck. Gently pull the crop from the breast, taking care not to rupture it. After it is free, grasp the esophagus and crop and pull until they come free from the body. Now grasp the trachea and pull it away from the body.

You are now ready to remove the neck. I find it easiest to use sharp poultry shears to snip the neck off at the bird's body. You can also use a knife to cut between the vertebrae and remove the neck.

ABOVE: Chickens that have been plucked and gutted cool off in water for at least two hours before they are wrapped.

There should now be a cavity remaining that has only a flap of skin (use this if you intend to stuff the breast as a flap to fold up over the stuffing and pinion to the top of the bird). Finally, spin the bird around to remove the nub and oil sacs lying on the top of the tail (some people don't bother with this step and instead cut the tail cleanly from the body. Some people claim that the glands, or "the Pope's Nose," give an off flavor to the meat, but this only occurs if the glands are intact during cooking. By removing the tail cleanly before cooking, you can avoid this step.

You have just completed the clean portion of gutting your bird. Now, the innards must be removed. Flip the bird on its back, and face the tail end toward you. To open the abdomen, make either a horizontal slit midway between the keel and the vent or a vertical slit from the keel down toward the vent. A horizontal cut will allow you to tuck the legs inside and roast the bird whole, while a vertical cut is used if the bird's legs are too short to tuck under the flap of skin, or it is going to be cut up into pieces.

After you have made your cut, grasp the vent with one hand and cut slowly between it and the tail upwards and around on both sides, toward the pelvic bones. Don't insert the knife too deeply, or you will perforate the intestine. Continue cutting around the vent until it is free from the outer skin, and then gently pull it away from the body.

Once the vent is free and you have made either your vertical or horizontal cut, it is time to draw or eviscerate the bird. Some people tug gently at the vent, pulling out the intestines in a long coil, until the point where they are attached to the gizzard. Alternatively, you can reach inside the bird, scraping the top of the keel with the back of your hand, and reaching all the way to the neck, gathering the organs and pulling back toward the cut you have made. Either way, the goal is to get everything out without soiling the inside of the bird. If the intestines break midway through, don't fret; just carefully rinse the inside of the bird out with fresh, cold water.

Make sure to remove the lungs—pink, sponge-like appendages that are pushed up against the ribcage—and eggs or testicles, depending on the sex and age of the bird. It might be a good idea to halve or quarter the first few birds you do, just to peer inside and see how well you've done gutting.

After the organs have been removed, the heart, liver, and gizzard can be salvaged. Pull the sac off the heart and trim it on the top. The liver should come cleanly away from the heart with a little tug, and you should be able to see the bright green gallbladder. Carefully cut this away with your knife and discard it. If you nick it and some bile spills onto the liver, rinse it well; bile is bitter and unappetizing. The gizzard makes excellent stock material after it has been emptied of its contents. Cut away the remnant intestine, leading to the stomach, and then, holding the gizzard in your hand, make a vertical cut along the edge. Now separate the red muscle from the yellow lining to butterfly it open. Inside will be a collection of hard objects, like stones, that the bird has used to grind the grain and other food items it has swallowed. Peel away the buttery, yellow lining from the muscle and discard. (As kids, we would often save the smooth stones in glass canning jars; a chicken's gizzard can do almost as well as a rock tumbler to polish the stones smooth.)

After your birds have been processed, they should be properly chilled and aged. Cooling the meat down to 40 degrees

Fahrenheit as quickly as possible will deter bacterial growth and prevent spoiling. On Fat Rooster Farm, the birds are rinsed after evisceration and then placed in a clean plastic or metal container filled with cold water and ice. They are left to cool for at least 30 minutes in this container before being transferred to another container of water and ice (a plastic or metal 33-gallon trash can purchased for this purpose works fine). The birds are left in these containers for at least four to eight hours to age. Alternatively, the birds can be loosely wrapped and kept for one or two days in the refrigerator to age. Aging tenderizes the meat by allowing the muscle to relax. If the birds are frozen immediately after killing, their meat will be tough and not as tasty as an aged bird's. Chapter Eight covers freezing and using your meat in further detail.

7

Using the Eggs You Produce

The cattle tracks on the frosted clover look like the snail paths in the neatly tilled rows of summer's gardens. Each cow's heavy hoof has roughly trampled ice crystals from the waning grasses into the still-warm soil, leaving emerald trails that eventually melt into the early morning's fog. They look the same, the cattle and the snail routes, but the smells, the sights, the chill air remind me that we are at the cusp of winter, that all is receding into sleep, senescence, torpor. Even the cattle are not duty-bound on their journey across the field, as the slugs were apt to be earlier, on their trail of a perfectly ripe tomato's scent. Instead, the cows and calves meander toward this corner and that, looking for the sweet green grass that has long gone, having been eaten carelessly, without a thought to fall's impending curtain call.

ABOVE: A chicken's track is left in the snow at Luna Bleu Farm. With the door open during the day, some hens walk from the bright greenhouse to the dark coop to lay their eggs.

The hens have begun molting their tired plumage for sturdier feathers to brave winter's chilly winds; chicks hatched on the sly in the corners of the barn have been released from their mother's overprotective gaze to search for food and shelter on their own. With the waning light come fewer eggs, fewer crowing cocks, no skirmishes over turf in the crowded henhouse. Every bird finds its place, leaving its neighbors be, waiting out the cold.

The farm's crops have been squirreled away in storage bins, earth cellars, freezers, and canning jars. Summer's helpers have departed on their separate journeys, like restless geese taking flight on the first of autumn's seductive winds calling to them with promises of what more there will be. Every year, it is a repeating pattern of green and gold, life, discovery, and growth. Then, gentle rest: the quiet darkness of winter, cold, and waiting.

Each of us who has not been beckoned away hunkers down, shutting out the cold and damp winds, patiently dreaming of the return of restlessness and caprice, the flutter and squawk of a mother hen protecting her brood: spring's swell, when the farm is rocked back to life, back to the soil and to the seed.

Nutritional Value of Eggs

Besides peddling wooden wagon loads of brightly colored gourds around the neighborhood for sale during the fall or mowing the lawn for the elderly couple across the street, my business ventures as a child also included chickens. We kept a calendar hung above the washing machine where my sister and I would carefully re-cord eggs collected and eggs sold. All of the money was stored in a piggy bank until the family vacation, at which point it was equally doled out to spend on whatever we liked. I once bought a stuffed, dried puffer fish in Florida that sat on my bookshelf, collecting dust along its spiny back until the cat knocked it to

ABOVE: A rooster of mixed heritage—"probably Rhode Island Red, Leghorn, and some others," according to farmer Ray Williams—roams the pasture at Back Beyond Farm.

the floor, where it shattered into a million little sea-creature pieces. I still have several of the miniature glass animals that I purchased at different vacation spots, housed in a special hardwood display case that Dad made me.

Aside from egg sales, the chickens gave us fresh eggs to eat on sometimes three out of seven family-sit-down-together breakfasts a week. Mom made poached, fried, or scrambled eggs, omelets, French toast, pancakes, and sausage egg casseroles. Back then, the egg had yet to be vilified as a health hazard, and we three girls were thin as rails and full of just as much energy as fluffy baby chicks.

The egg is an invaluable ingredient in the art of cooking. Not just for breakfast, eggs can be used to create anything from Hollandaise sauce to ice cream. Eggs contain nine essential amino acids, vitamin B-6, B-12, thiamin, folate, riboflavin, and many minerals; eggs are

ABOVE: A Black Star cross hen is one of 100 two-year-old hens laying eggs at Luna Bleu Farm. The chickens replaced a flock decimated by predators; that summer, farmer Tim Sanford shot about 17 skunks that were getting underneath an electric netted fence to kill chickens.

an exceptionally healthy food. Two arguments against the egg have unnecessarily hurt its reputation: high cholesterol content and the bacteria *Salmonella*.

Eggs are considered a high-cholesterol food item, and until recently, high cholesterol was implicated as the single most important culprit for heart disease. Cholesterol is manufactured naturally within the bodies of chickens and humans alike. It allows us to synthesize vitamin D from sunshine, essential for maintaining calcium and phosphorous uptake in the bloodstream and, ultimately, forming strong bones. Cholesterol is also necessary for the production of sex hormones. Recently, evidence surfaced that *saturated fats* were more important in heart disease than cholesterol. In fact, eggs contain less saturated fat and calories (2 grams and 75 calories for a large, 57-gram egg) than a small, lean burger has (268 calories, 7 grams saturated fat). Eggs are considerably less expensive to purchase than other sources of protein, and they are easier to produce on the homestead than steak or bacon. The United Nations Food and Agriculture Organization (FAO) rates the egg as a more efficient source of protein than the four other top sources of protein, higher in value than cow's milk, fish, beef, or soybeans.

If you're still concerned about eggs and cholesterol in your diet, you can leave out most of the yolks in your recipes, where the cholesterol is found. A good substitute for two eggs is one whole egg and two egg whites. Cook the remaining yolks and feed them to the family dog. You can also choose to raise your chickens using free-range pastured methods; recent studies have shown that both cholesterol and saturated fats are reduced in eggs from uncaged, pastured hens. Some tout fertile eggs or eggs from certain breeds as being lower in cholesterol than others. In truth, the way these birds are raised and their ability to forage for foods other than processed grains may be responsible for the lower cholesterol values in their eggs.

ABOVE: Rosie, a three-year-old Black Sex-Link hen, forages in the grass in the backyard at the home of Geoff Hansen and Nicola Smith. When Hansen bought three pullets from a local farmer for their new coop, Rosie was included as a mentor to teach them the egg-laying ropes.

Salmonellosis is caused by the bacteria *Salmonella*, a genus in which more than 2,000 species have been identified, but where only a few have caused the majority of salmonella outbreaks in chickens. The disease is most commonly grouped into four broad categories of disease: pullorum, or white diarrhea disease; typhoid; paratyphoid; and arizonosis. Of these, all are rare in North America, with the exception of paratyphoid. Specifically the antibiotic-resistant strain *Salmonella enteritidis* is of importance, because it can cause flu-like food poisoning symptoms in healthy adults, and it can be life threatening to the young, elderly, or immune compromised. The chance of contracting salmonellosis from eggs is still remote, about one in two million, though it can also be contracted from handling other infected foods. Cases of salmonellosis have been contracted from eating infected raw fish, sesame seeds, peanut butter, and undercooked red meat.

At one point or another, more than 70% of chickens are exposed to and infected by one of the 2,000 strains of *salmonella*. Birds that recover may then become carriers of the disease. Stress, overcrowding, or poor body condition from molting or lack of feed can stimulate an outbreak.

Salmonellosis is transmitted either through an egg laid by an infected hen or by contamination of the shell after the egg has been laid. Not all eggs laid by an infected hen are contaminated with the bacteria. Chicks may die within an infected egg at the end of incubation, or they can hatch and infect healthy chicks. The disease can also be spread through an unclean living environment: droppings, flies, rodents, wild birds, soiled shoes can all spread the bacteria.

The bacteria can cause inflammation of the bird's intestines, creating watery, milky, or mucous- and blood-tinged droppings. The disease can eventually lead to dehydration, emaciation, systemic infection (septicemia), and death. Antibiotics can be effective against the disease, though birds that recover can become carriers. Culling infected birds is the best management for controlling salmonellosis.

RIGHT: A bright-eyed, healthy bantam Mottled Japanese hen rests during the Northeastern Poultry Congress's annual Poultry Show at the Eastern States Exposition Center in West Springfield, Massachusetts.

To reduce your exposure to salmonellosis, keep your birds clean and dry, maintain clean nest boxes, collect eggs frequently, and discard any broken or cracked eggs. Alternatively, you can discard the yolks and use just the whites of the eggs, which contain antibacterial properties and don't harbor the bacteria. On the other hand, the yolk harbors the bulk of protein and amino acids, so you will be sacrificing the most nutritious part of the egg by using just the whites.

FRESHNESS AND SHORT-TERM STORAGE

It may be surprising to know that eggs can be safely stored without altering them for at least three months, provided they are clean and kept at a temperature of 40 to 45 degrees Fahrenheit and 70 percent humidity. Until used, they should be stored, small pointed-end down, in an egg carton in the refrigerator.

Provided that you will be collecting all the eggs from your hens at least once a day (and ideally two or three times daily), you won't need to worry about freshness. However, if your hens have free range on your farmstead, and you have stumbled upon a hidden spot where the hens have laid their eggs, there are several ways to be sure they are edible. The easiest method to determine freshness is to submerge the eggs in a pan of water. A fresh egg will stay horizontal on the bottom of the pan, but older eggs will list upwards, large end poking toward the top. Eggs that float to the surface should be discarded. The degree to which the egg floats is due to the air cell inside. When the egg is laid, there is little to no air cell, but as it cools and shrinks, a space forms at the large end, as the inner shell membrane pulls away from the outer shell membrane. Moisture escaping through the shell will also shrink the contents and increase the size of the air cell.

You can also candle an egg to determine the extent of its air cell. An air cell space larger than about ¼ inch indicates an old egg.

Of course, cracking the egg open will give you a good idea of its quality. A fresh egg will be almost odorless, having no sulfur smell. The white of the egg will be slightly cloudy and sticky, and the yolk will look like a firm, yellow moon. By comparison, an older egg that may not be good or safe to eat would have a clear watery albumen, watery yolk, and off smell.

A Barred Plymouth Rock hen and Araucana cross hen drink water in the pasture at Back Beyond Farm. While most of the chickens are on the farm for a year or two before their laying slows down and they are given away, the Araucana cross has been around since the farm was started in 2003.

Unless you are eating them within a week, cracked or heavily soiled eggs should be composted. As farmers, we are usually left with these eggs for our own consumption, and they are perfectly safe, providing they are clean and the inner membrane of the egg is still intact (it will look like parchment paper underneath the crack). If any liquid is coming out of the egg, it should be discarded.

LONG-TERM STORAGE

For longer-term storage, eggs may be frozen, canned as pickled eggs, or salted, methods for which will be covered later in this chapter. Other methods of storage were far more popular prior to the advent of refrigeration, such as oiling, thermostabilization, and storing in water glass.

OILING

Oiling eggs makes it possible to store eggs at higher temperatures than untreated eggs, and in clean cartons stored in a cool place, they keep for up to four months before developing an off flavor. Eggs are dipped in white mineral oil within 24 hours of being laid,

ABOVE: Whitey the White Plymouth Rock spreads her wings in the warmth of the sun in Rick Schluntz's and Carol Steingress's small in-town yard. The first time they had a very old and sick chicken, Steingress took her to the veterinarian to be euthanized. "I cried all the way home," she said. But the vet bill also taught her not to be as attached to the chickens.

then stored in closed cartons. Given that untreated eggs will safely store for up to three months in refrigeration with no oily mess to deal with, it hardly seems worth it to oil eggs, unless you are in an area with no cool place to keep your eggs.

WATER-GLASSING

Water glass, a chemical compound called sodium silicate and is dissolved in water, was a popular storage method in the early 1900s for whole eggs. It was used to store freshly laid eggs for up to six months or more, before whites began to get watery and yolks got runny, and eggs took on an off flavor. Reading the MSDS (Material Safety Data Sheet) for water glass, however, sounds like reading a brew recipe from Hecuba's cauldron in *Macbeth*. Classified as a moderate health concern, a fire hazard, and a material that is "harmful if swallowed or inhaled, causes severe irritation to eyes, skin and respiratory tract," water glass requires people working with it in the lab to use goggles and shield, lab coat and apron, vent hood, and proper gloves. The homesteader and small farmer are probably better off avoiding its use.

THERMOSTABILIZATION

Thermostabilization was often combined with oiling and is still a viable option if freezing or canning are not. Unlike eggs that have been oiled, thermostabilized eggs can also be used in recipes that require foaming, like cakes and meringues. The process destroys bacteria on the shell and seals the egg by coddling the first layer of white underneath the shell. Most of the egg remains uncooked, and the sticky albumin that was heated stays with the shell after it is cracked open, leaving a normal-looking white and yolk.

To thermostabilize, use freshly laid eggs that have been cooled to room temperature. Heat a pan of water to 130 degrees Fahrenheit—no hotter, or the eggs will cook, and no cooler or the bacteria will not be killed (use a thermometer). Using a steaming basket, submerge the eggs in the water for 15 minutes. Dry the eggs

RIGHT: A Black Star cross hen wanders outside at Luna Bleu Farm.

and store them in cartons. Storage conditions at 65 degrees will keep eggs fresh for up to two weeks, but eggs can be stored at 35 degrees for up to seven or eight months.

FREEZING

One of the easiest methods to keep surplus eggs for later use is freezing. Eggs should always be frozen raw or they will take on a rubbery, unappetizing texture. Eggs stored by this method can be used for up to one year and substituted in any recipe, from ice cream to quiche. Simply "scramble" the cracked eggs using a fork, slowly incorporating the raw yolks into the whites. Add salt or honey to the mix to preserve it, depending on intended use. To five medium eggs (about 1 cup of beaten eggs), add either ½ to 1 teaspoon salt or ½ tablespoon honey. The salt or honey acts as an emulsifier so that the yolks don't become pasty after freezing. Seal them in containers that are labeled with the number of beaten eggs and freeze for future use. I use one container with five eggs to make a quiche. The beaten eggs could also be frozen individually in ice cube trays for individual use.

PICKLING

If you want to preserve eggs for salads or hors d'oeuvres, as spicy deviled eggs, or you'd like to try a pickled food that's different from cucumber pickles or dilly beans, home-pickled eggs are the way to go.

Many people are afraid of pickled eggs. They envision two-gallon glass jars sitting on gas station counters with self-serve tongs ready to dip into a magenta- or oxidized orange–colored bath, in search of a rubbery oval, resembling food. Having nothing in common with the gas station variety, kids will often choose to eat a pickled egg over a hard-boiled one, and the nutritional value of the end product is just as good, if not better, given the added vitamin C component. They are packed full of flavor, and easier to eat than a hard-boiled egg. Pickled eggs are packed in vinegar and spice brine, so botulism, which thrives on high pH, has no way of surviving. The task of pickling eggs also tends to fall during a time when the memories of hundreds of pounds of tomatoes and zucchini, cucumbers, and beans, headed for the canning jar are distant and longed for. Pickled eggs will easily keep for six months after being sealed in canning jars with a boiling-water bath. Before pickling, whole eggs must be hard-boiled and shelled.

HARD-BOILING

The easiest way to hard boil eggs is to use eggs that are at least one week and no more than three weeks old. Fresher eggs will be almost impossible to peel without gouging out parts of the white along the way, while older eggs will have a larger air cell, causing the whites to thin and displace the yolk to the side, making the pickled egg appear deformed when it is sliced open. Pricking the wider end of the egg with a pin or adding a little vinegar in the boiling water also helps preserve the egg's original shape during boiling. If you boil eggs for too long, the yolks will turn an unappetizing khaki or green color.

LEFT: Two Rhode Island Red cockerels face off to see who will flinch first at Cloë Milek and Karl Hanson's home.

Place eggs to be hard-boiled in a gently boiling pot of water using a slotted spoon or ladle, taking care not to jostle or crowd the eggs to prevent cracking. Boil for no more than 10 to 15 minutes, then drain the pot and run ice cold water over the eggs for a couple of minutes. You can refrigerate them at this point or proceed with peeling. The cooler they are, the easier the peeling will be.

To peel, crack the shell by rolling the egg on a flat surface, pressing down firmly with the palm of your hand. Then tap each end, and proceed to peel the egg from its widest end. Unpickled, hard-boiled eggs will keep safely in the refrigerator for 10 to 14 days.

Fat Rooster Farm Pickled Eggs
For Brine

2 cups cider vinegar	2 tablespoons kosher salt
½ teaspoon dry mustard	½ teaspoon freshly ground black pepper
2 tablespoons whole coriander	1 bay leaf
1 teaspoon dried dill	1 teaspoon tarragon
1/8 teaspoon red pepper flakes	

For Jars

4 pint jars and their lids	16 to 25 hard-boiled eggs, depending on size
1 clove garlic, peeled	
1 small red pepper, for added heat	1 sprig dill

YIELD: Makes 4 pints

TIME REQUIRED: 35 minutes to hard-boil and peel eggs; 40 minutes to prepare and seal eggs in glass jars.

Hard-boil the eggs as described above. Peel the cooled eggs, making sure that no fragments of shell remain on the eggs.

While the eggs are boiling, wash the canning jars with hot, soapy water, rinsing them well. Fill a pot with water and sterilize the jars by submerging them in the water (it should cover the tops of the empty jars by about ¼ inch), and bring to a gentle boil. Keep the jars in the boiling water for at least 10 minutes. Sterilize the jar lids and rings in a smaller, shallow pan using the same technique.

Meanwhile, heat the brine in a saucepan on low heat. The brine should not boil, but be piping hot.

Remove the jars from their water bath (you can dump the water inside them out of the pot). Place the garnishes in the bottom of each jar and then stuff the hard-boiled eggs in the jars to the fill line. Using a canning funnel and glass measuring cup, fill the jars with brine to the fill line. Use a butter knife to displace air bubbles in the jars and then cover the jars with the canning seals and lids.

Return the jars to the pot and bring the water to a gentle boil. Process jars for 15 minutes. Remove the jars and let them cool completely on a flat surface. Make sure all jars have sealed; unsealed jars should be stored in the refrigerator. Ideally, the pickled eggs should be seasoned for at least three weeks before use. Pickled eggs can be stored safely after opening for at least three months in the refrigerator.

VARIATIONS: Add 1 teaspoon of turmeric or 1 whole peeled beet (remove before adding the hot liquid to the jars) to the brine to impart a yellow or reddish color to the eggs.

SALTING

Another popular way of preserving eggs, especially in Asian cooking, is salting. Impeccably clean eggs are brined in their shells in a salty solution for at least 12 days, and up to three to four weeks, at which point they can be stored at room temperature until use for months. Eggs are then boiled and used in various recipes, usually with rice, and in contrast to curries and other spicy Asian dishes.

Salty Eggs (Itlog Na Maalat, Haam Daan, or Kai-kem)

12 whole, raw eggs	1 cup kosher salt
4 tablespoons rice vinegar	1 tablespoon green peppercorns
6 cups water	1 gallon glass jar

The eggs you use for Salty Eggs should be absolutely clean and fresh. Use a stainless-steel scrubby to remove all debris from the shell (don't worry about removing the shell's protective cuticle).

Combine the remaining ingredients in a pan and heat until the salt dissolves. Cool to room temperature.

Gently place the eggs in the glass container and pour the cooled brine over them. Use a non-corrosive weight to submerge the eggs in the brine. Replace the jar's lid, and store it in a cool place for at least 12 days. Most authentic recipes recommend at least three to four weeks in the brine.

Before use, the eggs should be rinsed and then gently boiled in their shells for at least 12 minutes.

Salty eggs can be used in salads, to accompany spicy curry dishes, or in Haam Daan Ju Yoke Beng (Ground Pork with Salted Egg, below).

Ground Pork with Salted Egg (Haam Daan Ju Yoke Beng)
Serve this dish with white rice.

1 salted egg	1 ½ pounds ground pork
1 egg	1 tablespoon tamari or soy sauce
1 teaspoon honey	¼ teaspoon salt
1 teaspoon black pepper	1 tablespoon heavy cream
2 cups broccoli florets	½ cup green pepper, cut into long strips
½ cup carrots, cut into thin strips	

YIELD: Serves 4

TIME REQUIRED: 40 minutes

Mash the white of the salted egg separate from its yolk. Crumble the yolk and set it aside.

Using your hands, mix the pork, salted egg white, raw egg, tamari or soy, honey, salt, pepper, and cream. Let stand for 20 minutes. If serving with rice, prepare it while the pork mixture is resting.

Mix the yolk into the pork mixture with a fork. Form the mixture into a ball and place it in a bamboo steamer basket (if you don't own one already to accompany your wok, they are available at most Asian markets or online). Arrange the vegetables around the meat's edges in the basket. Steam until the meat is brown—about 30 minutes. Serve over the rice.

Breakfast Dishes

In the early part of the twentieth century, when breakfast was considered a pinnacle meal, families gathered to eat their fill and discuss the morning's chores and duties of the day. Sixty years later, after the advent of the refrigerator, the toaster, the microwave, Starbucks, and the demise of the small family farm, the traditional American breakfast is considered a rare and often elaborate occasion. Now, rushing the kids off to school and launching both parents on their journey to nine-to-five jobs requires simple, quick meals. Cold cereal, Pop-Tarts, and breakfast bars are the norm—hardly a proper way to break a more than eight-hour fast, especially if you're a child.

There are hundreds of ways that eggs can be prepared for breakfast. A great resource is Mark Bittman's *How to Cook Everything*, which devotes more than 20 pages and more than 100 egg recipes to the morning meal. Here are some personally tested and kid-friendly recipes.

Cracked Eggs

As a toddler, my son differentiated scrambled eggs from all other egg dishes as "cracked eggs." Some of the thrill of making "Cracked Eggs" may have just been the responsibility of choosing the brightly colored eggs from the basket to break into the mixing bowl. The texture of scrambled eggs is often more agreeable to children, and the addition of cream mellows the sharp egg flavor.

4 large eggs or 6 pullet or Bantam eggs	⅓ cup sour cream
I teaspoon dill weed	2 tablespoons heavy cream or whole milk
3 tablespoons butter	kosher salt and ground pepper to taste

YIELD: Serves 2 adults or 3 children

TIME REQUIRED: 20 minutes

Beat the eggs, sour cream, dill, and cream together using a whisk or egg beater until all ingredients are well blended.

Place a skillet on the stove over medium heat (I prefer a cast-iron skillet that has been well seasoned with use) for about 30 seconds. Add the butter and coat the skillet well. Pour in the egg mixture, turn the heat to low, and let it rest until a film has formed on the bottom of the pan. Take a spatula, tilt the skillet, and carefully push the egg toward the center of the skillet, allowing the liquid mixture to fill in the space left behind. Continue tilting the skillet back and forth until the center is relatively solid. Now the mixture can be gently stirred, and the eggs will break up into clumps. We like our eggs "over-hard," but they can be removed from the skillet at any time after forming clumps. Sprinkle with salt and freshly ground pepper, hot sauce, or ketchup. Serve immediately with toast or the eggs will "weep" liquid from the addition of sour cream and milk to the mix.

There are unlimited variations to this recipe. Enlist the help of a child to experiment and refine the recipe to his or her liking.

Poached Eggs

Poaching involves cooking eggs in gently boiling liquid until done. Special pans and inserts for poaching are made, and while they make the edges of the poached egg look a little prettier, they usually require butter or oil to keep the egg from sticking to the sides. The poachers are also too large for bantam eggs and too small for jumbo eggs, resulting in either overcooked eggs or whites that spill out of the pan's reservoirs. It is easy to crack the egg into a dish and then slide it into a regular pan of gently boiling liquid, to cook until done.

Poached Eggs with Mango Salsa

4 poached eggs	2 mangoes, peeled, seeded and coarsely sliced
¼ cup cilantro, coarsely chopped	
¼ to ½ cup maple syrup	½ cup red onion, coarsely chopped
3 tablespoons rice wine	juice of I lime
I medium tomato, coarsely chopped	3 tablespoons white vinegar
I jalapeño pepper, seeded and cored, or I teaspoon hot pepper oil	½ cup sweet pepper (green or other)
kosher salt and freshly ground pepper to taste	cilantro and chives for garnish

YIELD: Serves 2; you'll have leftover salsa

To make the salsa, place all ingredients in a food processor or blender and process until chunky but not fine. Adjust the seasonings (sweetness, heat, and salt) and refrigerate until use (will keep up to two weeks).

Poach the eggs and drain with basket or slotted spoon. Arrange each egg on a warmed plate with buttered toast or brown rice and a heaping dollop of salsa on the side. Sprinkle the eggs lightly with salt and pepper, garnish with cilantro and chives, and serve immediately.

Eggs Benedict

Delicate crêpes are used in this recipe to add to the creamy texture of this classic breakfast dish. If you don't have the time, substitute toast or English muffins for the crêpes.

For Crêpes

1 cup flour	2 eggs
½ cup milk	½ cup water
½ teaspoon salt	2 tablespoons melted butter

For Eggs

6 large eggs	1 avocado, peeled and sliced thinly
6 slices Canadian bacon	Hollandaise sauce (page 180)
fresh chopped parsley and minced chives	paprika
salt and pepper	peeled orange sections for garnish

YIELD: Serves 6

TIME REQUIRED: 1 hour 30 minutes, less if you have a helper

Slice the avocado and fry the Canadian bacon, and then make the crêpes. Keep the crêpes warm in the oven (170°F) while you are poaching the eggs. Poach the eggs, then keep them in the oven on a platter. Make the Hollandaise last, then assemble the dish and serve immediately.

Beverages

Recent health concerns regarding the potential for *salmonella* poisoning have removed raw eggs from their status as high–powered energy drink ingredients. Cookbooks written prior to the 1990s frequently included beverages with raw eggs without dire warnings accompanying them. While the addition of hot milk and heating will cook the eggs in these recipes to some extent, it may not be enough for all eaters. Because of the risks associated with eating undercooked eggs, no matter how remote their possibility in your fresh, home-raised eggs, the FDA suggests that children, the elderly, or those with compromised immune systems should make sure that the eggs are thoroughly cooked before eating them.

Eggnog

A non-alcoholic version of this beverage can be made easily by substituting 1 teaspoon vanilla extract for the liquor.

1 dozen eggs, separated	½ cup white sugar
½ cup maple syrup	1 cup good bourbon whiskey
1 cup cognac	1 quart heavy cream
2 cups whole milk	grated nutmeg for garnish

YIELD: Serves about 20

TIME REQUIRED: 5 to 6 hours

Beat the egg yolks, sugar, and syrup together until thick. Add the liquor (or vanilla if non-alcoholic) while continuing to beat slowly. Chill, covered, for several hours.

Before serving, combine the salt and egg whites and beat until nearly stiff, or until the whites form peaks that bend. Set aside.

Whip the cream until stiff. Fold the cream and milk into the chilled yolk mixture, then fold this into the egg whites. Cover and chill for one hour.

Pour the beverage into a punch bowl, top with the grated nutmeg, and serve in chilled cups.

Wassail

One of the biggest treats of the Wassail bowl, customarily passed from door to door in the poorer communities of England during Christmastime, was the floating pieces of bread swimming in the liquid, eaten with wishes of "Be Well"—which led to the expression, "drinking a toast."

7 apples, cored and sliced into ¼-inch wedges

1 cup maple syrup

750 milliliters sweet sherry

½ teaspoon ground cloves

1/8 teaspoon allspice

6 eggs, separated

5 slices whole-wheat bread, toasted, buttered, and quartered

1 ½ cups brown sugar

3 quarts light ale beer

6 thin quarter-sized slices fresh ginger

1 teaspoon nutmeg

2 cinnamon sticks

1 cup cognac, heated

YIELD: Serves 10

TIME REQUIRED: 1 hour

Sprinkle the apples with ½ cup brown sugar and bake in a preheated oven at 400°F for about 25 minutes.

Heat the ale, sherry, remaining brown sugar, maple syrup, and spices in a large saucepan until simmering (don't boil).

Beat the egg yolks until thick. Beat the whites separately until very stiff, then fold into the yolks.

Pour the ale mixture into the egg mixture slowly, beating hard so that the eggs don't cook.

Put the hot apples in a punch bowl. Add the ale-egg mixture and the heated cognac. Serve in warmed mugs, and float a piece of toast on top of each mug.

Sweets
Flourless Peanut Butter Cookies

Eggs act as a binding agent in this recipe, which lacks gluten. The cookies will be very soft when they first come out of the oven, so let them rest on the baking sheet before transferring them to a rack to cool.

3 eggs	3 cups peanut, almond, or cashew butter
2 cups white sugar	½ cup quick oats
2 tablespoons maple syrup	1/8 teaspoon vanilla extract
2 teaspoons baking soda	

YIELD: 2 dozen cookies
TIME REQUIRED: 1 hour

Preheat oven to 375°F. Lightly butter cookie sheets.

Beat the eggs, then add the remaining ingredients, mixing well until smooth. The dough should thicken as it is mixed. Roll into 1-inch balls and place about 2 inches apart on the cookie sheet. Flatten the tops gently with a fork, making a crisscross. Bake for 10 minutes, until golden brown.

Flink's Flan

Joey Flink was an apprentice at Fat Rooster Farm who was obsessed with flan. He took a *New York Times Cookbook* recipe and converted it so he could use his favorite sweeteners, honey and maple syrup. The summer was filled with flan trials, and at last, he became satisfied with this one. It's a wonderful dessert, but we had no problem eating it for breakfast, lunch, or as a snack.

¼ cup honey	2 ½ cups whole milk
4 eggs	1/3 cup dark maple syrup
1 teaspoon vanilla extract	1 teaspoon ground cinnamon
½ teaspoon ground nutmeg	½ teaspoon salt

YIELD: Serves 8
TIME REQUIRED: 1 hour

Preheat oven to 325°F. Place a pan with warm water in the oven, large enough to hold a 9 x 13-inch baking dish.

Warm the honey and spread it over the bottom of the baking dish. Heat the milk to scalding and set aside.

While the milk is heating, beat the eggs and combine with the syrup and spices. Whisk the milk into the egg mixture and pour into the baking dish.

Place the baking dish in the hot water bath and bake 50 minutes to 1 hour, until the mixture is not quite set (it will still jiggle in the middle).

Cool to room temperature, then refrigerate for at least 1 hour. Serve by inverting the sliced pieces of flan on serving dishes so that the honey drips over the entire portion.

Meringue Clouds

Instead of discarding the egg whites from your separated yolks called for in Hollandaise sauce, use them to make these light and fluffy sweets.

4 egg whites	¼ teaspoon kosher salt
1 teaspoon cream of tartar	1 ½ cups sugar
½ cup maple syrup	2 teaspoons vanilla extract

YIELD: 12 meringues

TIME REQUIRED: 45 minutes

Preheat the oven to 250°F. Butter and lightly flour baking sheets (alternatively, use waxed or parchment paper on the baking sheets).

Beat the egg whites until frothy. Add the salt and cream of tartar and beat until soft peaks form.

Gradually add the sugar and maple syrup while continuing to beat. Add the vanilla and beat until the mixture is glossy and stiff.

Using a spoon, drop the mixture on the baking sheets, 2 inches apart. Bake until firm and dry, but still white, about 30 to 40 minutes.

If desired, cool, put a scoop of ice cream or other filling between 2 meringues, and top with whipped cream.

Noodles and Breads

Spaetzle (Egg Dumplings)

Although these noodles can be served alone, they are also good accompanying Mean Rooster Stew (pages 190–192).

2 ½ cups flour	I teaspoon salt
¾ teaspoon baking powder	4 eggs, slightly beaten
I cup water	I cup butter
½ cup bread crumbs for garnish	

YIELD: Serves 4 to 6

TIME REQUIRED: 35 minutes

Combine the flour, salt, and baking powder in a large bowl. Make a well in the flour and add the eggs. Work them into the flour, slowly adding the water until the dough is moist.

Fill a pan with water and bring to a gentle boil. Place a colander over the water, and press the dough through it with a wooden spoon, a little at a time, into the boiling water. If the dough is too thick to force through the colander, add more water to it. Cook the dumplings for 3 minutes, or until they float to the surface. Don't crowd the noodles in the cooking water. Remove them with a slotted spoon to a warm dish when cooked.

Melt the butter in a saucepan. Add the breadcrumbs and brown until golden. Top the noodles with the crumbs, and serve hot.

Cheddar Corn Fritters

Fritters go well with pork dishes, chili, and other hearty stews. They can be served with warm sour cream or honey as dipping sauces.

2 cups fresh or frozen corn	4 eggs, separated
2 cups grated cheddar cheese	2 tablespoons fresh green chiles, roasted, peeled, seeded, and chopped
½ cup cornmeal	
2 teaspoons baking powder	2/3 cup flour
I teaspoon kosher salt	2 tablespoons maple syrup
	vegetable oil for frying

YIELD: Serves 4
TIME REQUIRED: 40 minutes

Combine the corn with the egg yolks, cheese, and green chiles. Stir in the cornmeal, flour, baking powder, maple syrup, and salt. Beat the egg whites in a metal bowl until they form peaks, then fold them into the fritter batter.

Put enough vegetable oil in a cast-iron skillet to come 2 or 3 inches up the sides, and heat to 375°F. Drop the batter by heaping teaspoons into the oil, taking care not to let the fritters crowd each other. Cook for about 3 minutes, then flip them over with a slotted spoon and cook for another 3 minutes, until golden brown. Remove them from the oil and drain on paper towels. Keep them warm in the oven, or serve immediately.

Spoon Bread

This great for when you want a quick bread to accompany chili or salty pork dishes. I make mine in a cast-iron skillet and serve it in the pan at the table.

3 cups whole milk	¾ cup cornmeal
2 tablespoons butter	I teaspoon kosher salt
2 eggs, separated	I teaspoon baking powder
green chiles, chopped fine (optional)	

YIELD: Serves 6
TIME REQUIRED: 45 minutes

Scald 2 cups of the milk in a medium saucepan. Mix the remaining milk with the cornmeal, add to the scalded milk, and cook over low heat for about 20 minutes. Cool slightly.

Beat the egg yolks and set aside. Beat the egg whites until stiff and set aside. Preheat the oven to 375°F.

Add the butter, salt, and beaten egg yolks to the cornmeal mixture. Add the baking powder and optional chiles and mix well. Fold in the egg whites. Turn the mixture into a greased casserole or cast-iron skillet and bake 30 minutes, until golden.

Buttermilk Doughnuts

Commercially available doughnuts have made the homemade versions somewhat of a rarity in these modern times. Growing up, doughnuts made on the stove were as common an item for breakfast on the weekends as French toast or pancakes. This recipe is adapted from the *New York Times Cookbook*.

4 ½ cups flour	¼ teaspoon freshly grated nutmeg
1 ½ teaspoons baking soda	1 ½ teaspoons cream of tartar
1 ½ teaspoons kosher salt	4 eggs
1 cup sugar	¼ cup melted butter
1 cups buttermilk (substitute ¾ cup whole-milk yogurt thinned with ¼ cup milk)	canola or safflower oil for frying

Yield: Makes about 3 dozen doughnuts

Time required: 40 minutes

Mix together the flour, nutmeg, baking soda, cream of tartar, and salt.

Beat the eggs until thick, and gradually beat in the sugar. Add the melted butter and buttermilk, then add the flour mixture. Mix well and chill for about 30 minutes.

Turn the dough out on a well-floured board. Roll to ¼ inch thick, then cut with a floured cutter.

Fry a few doughnuts at a time in the hot oil (use a cast iron Dutch oven or frying pan filled about 3 inches deep with the oil and heat it to 375°F) for about 3 minutes on each side, or until brown.

Sauces, Condiments, and Dressings

Hollandaise Sauce

The hardest part about making this sauce is having it break, or curdle. If this happens, just whisk a little warm water into the finished sauce and serve immediately.

I stick (¼ cup) softened butter	3 large egg yolks
2 tablespoons lemon juice, preferably fresh	I/8 teaspoon cayenne pepper—salt to taste

YIELD: Makes about 1 cup

TIME REQUIRED: 25 minutes

Combine the egg yolks and water in the top of a double boiler and whisk over hot (not boiling) water until fluffy.

Slowly add pieces of the butter, beating with the whisk continuously as it melts. Add the salt and lemon juice. Continue beating until all ingredients have been added and the mixture is thick. Stir in the cayenne, and serve the sauce immediately.

Mayonnaise

I whole egg	I egg yolk
½ teaspoon kosher salt	2 teaspoons lemon juice
I ½ cups corn, canola, or safflower oil	

YIELD: Makes about 2 cups

Beat the eggs, then mix with the salt and lemon juice in a blender or food processor. Turn to high, and slowly drizzle the oil into the blender jar until the mixture is thick. Pour mayonnaise into a storage container and add fresh herbs, chopped onions, cayenne pepper, or other spices, as desired. The mayonnaise will keep in the refrigerator for one week.

Tartar Sauce

I cup mayonnaise	I tablespoon parsley, chopped
I tablespoon chives, chopped	¼ cup pickle relish

YIELD: Makes about 1 ¼ cups

Combine all ingredients and blend well. Add chopped garlic, capers, and tarragon if desired.

Five-week-old "Barnyard Classic" chicks hatched out in an incubator huddle together in the coop.

> **"**The first rule to remember if you plan on raising chickens for meat is never to name a bird you intend to eat! Either you won't be able to "do it" when the time comes, or that beautiful roast chicken will sit on the table while you and the kids sit around with tears in your eyes.**"**
>
> **–RICK AND GAIL LUTTMAN IN *CHICKENS IN YOUR BACKYARD*, 1976**

⑧

Using Your Chicken Meat

How to Store Your Meat

Birds can be frozen whole or cut up into halves or quarters. If you plan on storing them for longer than one or two months, and you don't have the ability to cryo-vac the meat, freezing them whole will reduce the amount of freezer burn (less surface area within the container). Prior to freezing, be sure that the birds are dry so that no water remains inside the body cavity. Storing the birds in freezer bags works well; double-wrapping in freezer paper will further reduce the potential for freezer burn.

After you have butchered and cleaned your chicken meat, it is important to age it. The texture of a freshly killed chicken will be tougher than if you wait to eat it at least until the next day; the muscle needs to soften and bacteria will actually mellow the meat. The bird should be stored at less than 40 degrees Fahrenheit at all times during aging. The best way to do this is to chill it in ice-cold water (less than 40 degrees) for at least 6 hours (I prefer to chill and soak the birds for 12 to 18 hours).

After aging, the birds should be thoroughly dried. At this point, they can be kept for up to four days in the refrigerator, or frozen and stored for 6 to 12 months (I have eaten chickens that have been in the freezer for up to two years, but the meat loses quality and flavor). Remember that freezing does not kill bacteria, so the birds should be impeccably clean before storing in the freezer. A whole chicken will store longer than a cut-up one (13 months versus 9 months).

FREEZING

Everyone has a favorite storage technique for packaging up chicken for the freezer. Cryo-vac machines do the best job of removing as much air from the package, thereby reducing potential for freezer burn and bacterial growth, but plastic bags designed for freezer storage work fine, providing as much air as possible has been forced out. One disadvantage is that the plastic can easily be punctured if the packages are jostled about in the freezer too much.

Double-wrapping the birds with freezer paper is tedious, but there is less chance of freezer burn or puncture of the package as it is jostled in the freezer. Some home preservers advocate first wrapping the chickens with foil or plastic wrap and then finishing them in freezer paper. If you are planning on selling the chickens, a clear wrap is better than freezer paper; consumers like to see what they're purchasing.

Cooked chicken can be frozen, but it will only keep for three to six months before it loses quality and flavor. Broth and stock is easily stored in reused containers such as yogurt or cottage cheese tubs. While the quality will degrade within three months, it will store for up to six months in the freezer.

If you are relying on your freezer to store the family's meat for the year, and you live in an area prone to power outages, you should have a backup plan. Having a generator on hand is probably more economical than searching for a source of dry ice during a raging storm (besides, you can use it to run the incubator you've set with eggs to replace the laying hens).

ABOVE: Chickens raised for their meat are part of Sunrise Farm's operation in Hartford, Vermont.

At the very least, check your freezer at least every two days. There is not much more depressing than opening a freezer full of spoiled meat that you have patiently and carefully raised and stored away. In case you have had a loss of power, don't panic. A freezer full of food will keep fine for up to two days, provided it is well stocked and you keep the lid closed. Even chicken that has begun to thaw is safe to refreeze, as long as ice crystals remain in the meat. In fact, the USDA maintains that thawed chicken that has been kept at 40 degrees Fahrenheit or colder for up to two days can safely be refrozen. Even chicken that has completely thawed for longer can be cooked and refrozen. Any meat with an off flavor or color should be composted.

Thawing your meat for use can be accomplished slowly in the refrigerator the day before, on the counter at a temperature no higher than about 70 degrees in about 10 hours, or in a brine (page 188) or cool water bath in about 2 hours. Of course, you can always use a microwave to thaw the meat within minutes.

CANNING

If refrigeration is an issue, canning the meat is an option. Unfortunately, meat is a low-acid food and may contain harmful bacteria, which can cause botulism. At 240 degrees Fahrenheit, these bacteria are killed, but only a pressure canner can safely achieve this temperature within the canning jar.

Chicken meat for canning can be prepared by raw- or hot-packing. Deboning at least the breast and thighs before you can them is more practical; it saves space, and the end product is easier to use. For step-by-step canning procedures, *Stocking Up* (see Recommended Reading) is invaluable.

Although the risk of spoilage is higher, curing and smoking poultry meat is also possible. Home-raised and -processed birds are excellent to use because they are fresher than any that could be bought in the store. Unfortunately, the only safe way to guard against potentially dangerous bacterial contamination of the meat is to use a commercial cure that contains nitrates (see Appendix Four). Washington State Cooperative Extension (see Appendix Three) has an excellent website with information on curing and smoking poultry, and there are numerous other sites with similar recipes and instructions, including those that cure and smoke without using nitrates.

Cutting Up a Whole Chicken

Our farm offers a CSA (community-supported agriculture) arrangement where people purchase "shares" before the growing season begins, paying us to purchase seeds, potting soil, chicks, grain, or what-have-you. In turn, the members receive the farm offerings at a discounted value. As part of the offering, whole organic chickens are included. One member, a neophyte to the CSA idea but interested in doing her part to support local agriculture, was perplexed. Would the chicken still have feathers on it? It suddenly occurred to me that in our world of bagged chicken wings, boneless breasts, and packages of skinless thighs, unless the chicken is to be roasted whole, much of America no longer knows how to deal with a whole chicken (and no, the chicken does not still have its feathers).

The good thing about raising and butchering your own chicken meat is all the leftover goodies. Chicken bones are great to use for making chicken stock, as are necks, gizzards, hearts, and livers, and cutting up the bird is really not difficult. Even the trimmed fat can be rendered and saved for cooking.

To break down a whole chicken, you'll need sharp poultry shears or a good knife, such as a Santuko, or another sharp, heavy knife. You'll also need a sturdy cutting board, preferably one designated for meat cutting, some patience, and a little confidence that with practice, you will be able to do this easily.

1. Place the whole chicken on its back, and cut through the skin joining the leg to the body. Bend the leg backward, until you feel the leg joint pop free. Then flip the bird over and, from the midline, slice down, starting at the attached end of the leg and cutting toward the tail, freeing the "oyster" (the small pocket of meat on the back), and through the separated joint to cut off the leg (the cut will look like a backwards "C"). Repeat this process on the other side.

2. Divide the leg into two pieces at the thigh by bending the leg backwards and slicing through the meat and between the joint.

3. Cut the wing from the body in the same manner used to free the legs. Trim the tip off the wing at the first joint (you can reserve this small piece of meat for stock). The wing can also be trimmed in two, as the leg was, to make buffalo-style wings.

4. Free the back from the breast by standing the bird upright and slicing or cutting through the skin from the back to front, close to the ribs. Then bend the back and cut through to where the neck was. Use the back in stock for soup.

5. To split the breast, poultry shears work best, cutting along the midline. A sharp knife can also be used, but first lay the breast skin-side up, and with the palm of your hand press firmly down onto the cutting board until you hear the breast crack (this technique is also used to butterfly poultry). Use the knife to divide the breast in two by slicing through the cartilage as close to the midline as possible.

Brining

Submerging poultry in a solution of salt, sugar, and spices for a period of time infuses the meat with exotic flavors and plumps its flesh, making it juicy. It is also an easy way to thaw frozen birds gently, rather than leaving them on the counter or the refrigerator to dry out while they thaw. Remember to reduce the amount of salt used in the final recipe if you brine your bird, or the end product will prove too salty.

½ cup kosher salt	¾ cup sugar
½ cup maple syrup	3 whole bay leaves
I tablespoon black peppercorns	I tablespoon mustard seeds
2 sprigs fresh sage or I tablespoon dried	I sprig fresh rosemary or I tablespoon dried
2 sprigs fresh tarragon or I tablespoon dried	I gallon warm water plus more to cover bird

YIELD: Makes enough brine for a 4 to 6-pound chicken

In a large, non-reactive pot, dissolve the salt and sugar in the warm water. Add the remaining ingredients, and stir the mixture well. Submerge the chicken, frozen or thawed, in the liquid, cover, and place in the refrigerator or a cool place (less than 40°F) for 2 to 8 hours.

NOTE: Additional ingredients such as dill, thyme, lemons, juniper berries, allspice, fennel, star anise, and whole cardamom can be added to the brine, or it can be as simple as a salt-and-sugar mixture.

Chicken Stock

The definitions between stocks, broths, and soups are often muddied. Alan Davidson, in his tome *The Penguin Companion to Food* (2002), distinguishes between the three using these simple guidelines: a stock generally implies liquid in which meat or vegetables have been cooked and will be used to create further dishes. Broths can be consumed as is, or used to create soups, which are generally regarded as finished dishes. Whether I have removed the meat from the chicken's carcass, or cooked it whole on the grill or in the oven, the remains always wind up in a pot to be made into stock for later use as broths or soups.

Stock from a Cooked, Whole Chicken

6 quarts water	¼ cup dry red wine
carcass and leftover pieces, such as wings and legs, from roasted or grilled chicken	½ pound kale, mustard greens, spinach, chard, or other leafy greens
2 carrots, scrubbed, unpeeled, cut into 3-inch chunks	1 large (6 to 8-ounce) sweet onion, unpeeled, stuck with 2 whole cloves
2 shallots, unpeeled	2 cloves garlic, unpeeled
2 large celery stalks, leaves included	½ cup fresh parsley
½ teaspoon ground black pepper	2 bay leaves

YIELD: Makes about 4 cups stock

TIME REQUIRED: 3 hours

Submerge the chicken carcass and remaining ingredients in water and wine. Bring to a simmer, and skim off any foam from cooking surface. Don't let mixture boil.

After the liquid has reduced by about one-third (about 2 hours), remove from heat and strain through a fine sieve. Stock can be used immediately or frozen for up to three months (it will safely freeze for longer, but the quality of the stock will quickly deteriorate after three months).

NOTE: Other vegetables can be added to make the stock, such as turnips, potatoes, tomatoes, and parsnips. All of these tend to sweeten the stock a bit. Salt has been omitted here, but can be added to taste after the stock is finished.

Stock Using Uncooked Chicken

Here, the chicken can be cooked whole or with the meat removed. A stock using whole old hens or roosters has a much stronger chicken flavor than one made using the uncooked carcass left from deboning the whole bird.

Proceed as for making stock from a cooked chicken. The liquid will need about 1 to 2 additional hours to cook down, and after straining it should cool and be skimmed of excess fat before using or freezing.

Appetizers

Chicken Livers with Ramps, Raisins, and Rhubarb

Ramps or wild leeks are native to the east as far south as Georgia and west to some parts of Illinois. Their blankets of green leaves in rich forested areas are one of the harbingers of spring and a welcome addition to onions and garlic that have been stored for several months in the earth cellar. If ramps are unavailable, substitute leeks, mild onions, or shallots. This recipe is adapted from my friend Scott Woolsey's, who is manager for Killdeer Farm Stand in Norwich, Vermont.

1 ½ cups rhubarb, chopped	1 cup raisins
¼ cup honey	2 tablespoons butter
¼ cup ramps, roots trimmed, leaves and bulbs chopped	12 ounces chicken livers, trimmed and cut in half
2 cups dry white wine	kosher salt and fresh ground pepper to taste
fresh thyme and sage, minced, about 1 tablespoon total	

YIELD: Serves 8

TIME REQUIRED: 20 minutes

Toss the chopped rhubarb, raisins, and honey in a bowl. Melt the butter in a sauté pan over medium-high heat. Add the ramps and cook until fragrant, about 2 minutes.

Add the chicken livers and sauté for 2 minutes on each side; remove and let cool. Add the wine to the pan and reduce by half.

Add rhubarb mixture to pan and simmer until the rhubarb is soft.

Return the livers to the pan. Add salt, pepper, and herbs and cook for 2 minutes, until livers are cooked through.

Serve on toast garnished with chive or violet flowers.

Rumaki

In the *New York Times Cookbook*, this appetizer is described as being almost as popular as pizza pie in metropolitan America. While liver in general has fallen out of favor, I encourage you to try your fresh home-grown chicken livers. This dish is a great way to experience them for the first time.

½ pound chicken livers, trimmed	I can water chestnuts
9 slices bacon, cut in half	9 ramps, leaves on, sliced in half (substitute green onions)
½ cup tamari	
½ teaspoon curry powder	¼ teaspoon ground ginger

YIELD: Serves 6

TIME REQUIRED: 3o minutes

Slice the chicken livers into thirds and fold each piece over a water chestnut.

Wrap a strip of bacon and a ramp around the stuffed liver and pin it with a toothpick.

Mix the remaining ingredients and marinate the livers for 1 to 3 hours in the refrigerator, turning occasionally.

Remove the livers from the marinade, and broil in an oven, turning frequently, until the bacon is thoroughly cooked, about 5 minutes.

Chicken Sides and Mains

Mean Rooster Stew (Coq au Vin)

I have just completed a frantic telephone call with a French chef at a top restaurant in a Vermont resort town. He can't procure the necessary ingredients for his signature dish, Coq au Vin. In his native France, it calls for a rooster, preferably two or three years of age complete with its blood, feet, comb, and wattles. Our governing laws in the United States prohibit the sale of many ingredients such as blood and certain animal parts under the premise that such foods carry a higher risk of disease and illness potential if consumed by humans. My own American reaction to this chef's request is a hardy "Yuck."

In fact, many European chefs will tell you that they never use a young chicken or stewing hen in this dish, as the flavor turns out bland, and the fat from plump meat birds or hens gives the dish an off flavor. Here's my own recipe for Coq au Vin, toned down for home cooks, but still a great way to do away with unwanted roosters.

FIRST:

1 chicken, brined, preferably a rooster, then cut up as for frying	3 cups red wine such as Burgundy, Beaujolais, or Chianti
1 large sweet onion, diced	1 carrot, diced
1 stalk celery, cut into slices	1 tablespoon black peppercorns

Combine the ingredients in a non-reactive container and marinate overnight in the refrigerator.

SECOND:

¼ pound bacon, cut into rectangles ¼ inch across and 1 inch long	2 tablespoons butter
	½ teaspoon black pepper
½ teaspoon kosher salt (omit salt if the chicken has been brined)	½ cup flour
strained marinade	1 tablespoon minced garlic
½ tablespoon tomato paste	2 bay leaves
¼ teaspoon thyme	

Remove the chicken from the marinade and set aside. Strain the liquid and reserve; discard the remaining solids.

Dry the chicken pieces and lightly coat with flour. Shake excess flour from the chicken, and season with salt and pepper. Return the bacon to the pan, add the chicken, and brown the pieces on all sides.

Add the marinade to the pan with the chicken and bacon. Stir in the tomato paste, garlic, and herbs and bring to a simmer. Cover and cook slowly for 45 minutes to an hour, until the chicken is tender and juices run clear when pricked with a fork. Remove the chicken to a side dish and reduce the liquid in the pan to about half the original volume. Adjust the seasoning, and remove the bay leaves.

While the chicken is cooking, prepare the mushrooms and onions.

THIRD:

½ pound mushrooms, sliced in half	12 to 15 small white onions, such as cipollini
2 tablespoons butter	
pinch of salt	pinch of sugar

Heat the butter in a skillet until foamy. Add the mushrooms and sauté until golden brown. Set aside.

Add the onions, sugar, and salt to the melted butter. Add enough liquid to just cover the onions, bring to a boil, and then reduce the heat to simmering. Cook until the liquid has evaporated and the onions are tender.

ASSEMBLING THE DISH:

Return the chicken, mushrooms, and onions to the reduced wine sauce and heat thoroughly. Adjust the seasoning one final time, and serve the dish with buttered noodles such as Spaetzle or Egg Noodles (page 176).

Garlic Chicken Soup

This is a transitional soup that uses the last of your winter's larder intermingled with spring's first offerings. Be sure and take advantage of fresh offerings, such as stinging nettles, wild ramps, dandelion greens, or tender chive shoots.

I chicken, 3 to 4 pounds, cut into pieces, preferably brined	8 cups water
	¼ cup chopped lovage leaves
I cup sliced white or yellow onion	
	I ½ cups sliced carrots
I ½ cups rice	
	3 cups dandelion leaves, spinach, or Swiss chard
½ cup chopped parsley	
2 cups violet leaves	2 teaspoons kosher salt
½ teaspoon freshly ground pepper	½ cup grated cheddar cheese

YIELD: Serves 4

TIME REQUIRED: 40 minutes

Put the chicken pieces in a soup pot with the water, onion, and lovage and simmer the broth for at least 2 hours, or until the meat is readily picked off the bone.

Debone the meat and chop it coarsely. Strain the stock and return the meat to the broth.

Add the rice and carrots, and simmer for 30 to 40 minutes. If the soup is too thick, add more water.

While the soup simmers, wash and coarsely chop the greens and violet leaves. Just before the carrots and rice are tender, add the chopped vegetables, all but 2 tablespoons of parsley, and season with salt and pepper to taste.

Serve in bowls garnished with the remaining parsley and grated cheese.

Beer Can Chicken

The first time I ever heard of such a thing, my reaction was, "Why on earth go through the trouble of stuffing a beer can into a chicken?" The answer is simple: First, it's a great conversation starter at a barbecue; second, it's a great way to use up the cheap beer that someone left the last time you threw a party; and lastly, the chicken tastes great! The beer acts as an internal steamer, filling the bird's cavity with fragrant steam, all the while allowing the outside to cook slowly, as it soaks up the smoky tones of the grill.

2 whole 2 to 3-pound chickens (that have been previously brined)	2 inexpensive cans of beer, such as Miller or Budweiser
4 tablespoons olive oil	¼ cup whole basil leaves
freshly ground black pepper	

YIELD: Serves 4

TIME REQUIRED: 1 hour

I like using a gas grill instead of a charcoal-fired one. The temperature of the flame is more easily regulated, and you don't run the danger of running out of heat halfway through if you haven't used enough charcoal. You can parboil the whole birds for about 20 minutes before placing them on the grill. While the meat will not take on as much flavor, there is less chance that the birds will not be completely cooked.

If the birds have been brined, slather them with the olive oil and sprinkle liberally with black pepper. Unbrined birds should also be sprinkled lightly with kosher salt.

Open the beer and pour out about one-third of the can. Take a can opener and make three holes around the top of the lid, so more steam can escape during cooking.

Standing the chicken on end, legs pointing down, seat the can inside the cavity. The bird should be stable enough to stand upright.

Stuff the neck opening with the basil leaves.

Place the birds on the grill. If the lid to the grill does not shut, wrap the edges with aluminum foil to create a tent.

Cook for 40 to 55 minutes, until the chickens take on a golden color and a leg can be easily bent back, away from the beer can.

Broiled Chinese Black Fowl in Dark Soy Sauce

I have already determined that I don't have the fortitude to do in my cute, fuzzy Silkies, also known as Chinese Black Fowl. However, the roosters can be obnoxious, so if you find yourself with too many, and no one is in need of another pet rooster, here's an option from my sister in West Virginia.

2 tablespoons canola oil	I large Spanish onion, diced
5 cloves garlic, smashed	¼ cup Chinese bean paste
2 whole 2 to 2 ½-pound chickens, head and feet discarded, cleaned and rinsed	1-inch-thick piece of unpeeled ginger, cut into 4 pieces
1-inch-thick piece unpeeled fresh or frozen galangal, smashed	2 Thai bird chiles
½ cup wolfberries, or ¼ cup finely diced dried apricots	¼ cup diced jujubes (also known as hong zao or red dates) or 1/8 cup regular dried brown dates, diced
3 cups dark soy sauce	I cup cola (like Pepsi or Coke)
I star anise pod	2 cardamom pods
2 cinnamon sticks	I clove

YIELD: Serves 4

NOTE: Bean paste, chickens, galangal, chiles, wolfberries, jujubes, and dark soy sauce can be bought in Chinese food stores.

Place a large, heavy-bottomed saucepan or Dutch oven over medium heat, and add the oil. When the oil is hot, add the onion, garlic, and bean paste, and sauté until the onion is tender.

Add the chickens, ginger, galangal, chiles, wolfberries (or apricots), jujubes or dates, soy sauce, and cola. Stir, and add star anise, cardamom, cinnamon, clove, and just enough water to cover the chickens. Bring to a boil, then reduce heat to low. Simmer, covered, for 45 minutes. Remove the chickens, cut into serving pieces, and serve with broth ladled over the chicken and steamed rice.

If you want a thicker sauce, remove the chickens. Cook the broth over high heat until it becomes syrupy and glossy, about 20 minutes. If too salty, add a little water. Cut the chickens into serving portions, add to the sauce, heat through, and serve.

Chicken Sausage

For whatever reason, be it health-related or the fact that Americans now eat more chicken than any other type of meat, chicken sausage is all the rage. Supplies for sausage making are easily found on the Web, and grinding and stuffing attachments for your home food processor or blender can be inexpensively purchased. Making sausage out of tough laying hen or rooster meat is an alternative to stock and soup. The following recipe is adapted from Perry and Reavis's *Home Sausage Making* (2003). The sausage can be stuffed into casings and pan-fried, or it can be made into patties (you may need to add a beaten egg to bind the meat together into patties). The sausage can be used within three days or frozen for up to two months.

hog or sheep sausage casings (see Appendix Four)	2 pounds cubed raw chicken meat with skin, preferably from old laying hens or mean roosters
3 teaspoons kosher salt	
½ teaspoon cayenne pepper	2 teaspoons fresh ground pepper
I teaspoon dried sage, ground	½ teaspoon ground ginger
½ teaspoon dried thyme leaves	½ teaspoon dried rosemary, ground
	I teaspoon dried marjoram or oregano, ground

If you are using casings, prepare them for use. Snip off about a 4-foot section and run cold water over it to rinse off the salt that it was packed in. Then hold the opening of the casing under the faucet nozzle (I actually slide it up over the nozzle) and run water through the casing. Soak the casing in a fresh bowl of cold water with a little vinegar added (about 1 tablespoon for each cup of water) and leave it there until it is ready to be used.

To make the sausage, cut up the meat, fat, and skin into 1-inch cubes and put it in the freezer for about 40 minutes to chill. This makes grinding the meat easier.

Measure out all of the seasonings into a separate bowl, and set them aside. If you are using an attachment to your mixer or a food processor, grind the meat first, then mix in the spices by hand afterwards and mix well. If you are using an electric grinder, mix the meat and spices together well, then grind the meat. I usually like to grind the meat coarsely once, then put a finer grinding wheel on as the meat fills the casing.

After the casings are filled, prick the entire length of it with a pin to make air pockets. Then twist the casings off into 2- or 3-inch lengths. Cut the links with a knife and use them immediately or store in the refrigerator or freezer. Cook the sausage to an internal temperature of 165°F.

Jillian Noble, 9, of Dracut, Massachusetts, waits her turn with her Modern Game chicken during the Junior Showmanship Competition at the Northeastern Poultry Congress's annual Poultry Show at the Eastern States Exposition Center in West Springfield, Massachusetts. Noble has been showing chickens for four years.

9

Showing Chickens

"In the summer of 1990, I got a phone call. It was my mentor, Dick Holmes, a master breeder of poultry, saying that Clint Grimes couldn't care for his own birds any longer. He had had a bad turn of health and couldn't leave all the work to his wife, and someone needed to go there 'tomorrow' to pick his birds up. I was elected to make the 200-mile trip, and I arrived early that Sunday morning with every crate I could find lashed onto my pickup truck.

"It took four hours for us to catch and load them, but we found time to go over the birds he had and cull out those he felt were unsuitable for breeding stock to preserve his magnificent line.

"I had managed not to cull out any good ones, but then he grabbed the birds and told me that now he wanted to go over a few things. He picked each bird up and dug through its feathers here and there, never saying a word. Master breeders have mastered all of the basic knowledge regarding breeding for type and color. They focus their efforts on fine points and refining the strain of the breed they're working with. Clint saw the faults in those birds, but he wasn't going to let me in on what they were. He had me, hook, line and sinker, and it took me years to realize that the reason he never told me exactly what he was looking for was so I could figure out the finer points myself."

—DON SCHRIDER, COMMUNICATIONS DIRECTOR, AMERICAN
LIVESTOCK BREEDS CONSERVANCY

For a real taste of chicken fever, you need to attend a poultry show. At first, the sheer number of chicken breeds that exist will overwhelm you. Big, small, smooth-feathered, rough-feathered, calm and collected, jittery and frantic, funny-looking, naked, covered in fuzz, anything is possible. Then, the cacophony of sounds hits you. Loud crows, like trumpets, crows that are so ear-splitting and high

that you wish you had earplugs. Hens tut-tutting about, looking for a place in their display cage to lay the inevitable egg, squabbles, squirmishes, and brawls between neighbors. It's hard not to want every single breed you encounter.

And then there are the exhibitors themselves. Fussing over their chickens, primping this crest of feathers or that toothed comb, tapping the bird here or there to set up a pose typical of the breed, offering their birds food just at the right time to entice their birds to action for the poultry judge, willing to tell you anything you'd like to know about the best breed in the world—the one that they are showing.

LEFT: A Bantam Belgian D'Uccle rooster eats while waiting for the judging.

Why Exhibit Poultry?

The advent of exhibiting poultry may have started out as simply as five guys at a bar, asking the barkeep to decide whose hen was the best, but in the end, they became the trade shows of the poultry industry. Breeders exhibited to advertise their best birds, drumming up business for their lines of poultry and interest in their breeds. E. B. Thompson showed his champion Barred Plymouth Rocks at Madison Square Garden in the early 1900s, and for years no one won a prize in that category but Thompson. If you wanted Barred Plymouth Rocks, you went to him.

Today, poultry shows are a way to keep diversity in poultry breeds alive. In this accelerated era of information, preserving poultry genetics is essential. Our poultry production in the past 50 years has been completely dismantled and redesigned and may not be sustainable enough to take us through the next 200 years. Better-quality feeds have emphasized production of hybrids that don't need to forage on pasture to feed themselves. While these hybrids are faster-growing meat chickens and more productive laying

BELOW: Judge Jerry Yeaw of Scituate, Rhode Island, inspects a Bantam Old English white pullet while clerk Kyle Yacobucci of Palatine Bridge, New York, takes notes. Yeaw said he's been involved in the showing, breeding, and judging of chickens for nearly 70 years.

🐔 Chicken Scratch

In 2000, my husband and I traveled to Columbus, Ohio, to see the largest, most popular poultry show in the United States. We bought a black Minorca cockerel; on the 800-mile drive home, explaining his presence in the back seat of the Toyota sedan to the patrolman who pulled me over for speeding got me out of a ticket.

For a truly spectacular show, make the trip. It's held each year in early November, by the Ohio State Poultry Breeder's Association (www.ohionational.org). If you can't make the Columbus show, there are also shows on the East and West Coasts. Just Google "poultry shows," and you'll find locations and dates for shows in your area. If you decide to enter a show, you'll need to choose an event and apply for the show schedule. This will give you an entry form and all the pertinent information you need to be on time for your event.

hens, they are becoming less genetically diverse and therefore more prone to disease. Thinner gastrointestinal walls, designed to absorb nutrients more efficiently, have also left them more prone to bacterial uptake and danger of infection. Confinement to cages for laying and to open-air barns for meat production are the present technology, but we need the diversity of poultry breeds to ensure our ability to switch our management practices if we need to.

Not only is breeding and exhibiting poultry an important component of preserving a diverse variety of breeds, it is also an ideal way to introduce young people to the joys of keeping chickens. Exhibiting teaches responsibility for living creatures and attention to detail.

LEFT: Calvin Rambacher, 5, cuddles with Little Smurf, a Blue Andalusian hen, during the showmanship competition. It was Rambacher's second time showing his chicken.

Becoming a Good Showperson

If you'd like to enter the world of exhibiting poultry, you should stick to some simple rules.

· Concentrate on a single breed—don't catch chicken fever. Each breed has different management requirements and conditioning strategies. If you spread yourself too thinly, you'll end up with a bunch of barnyard classics—glorified mutts that may be cute to look at, but don't serve much purpose in preserving the diversity of chicken breeds.

· Use the *Standard of Perfection* to select desirable traits for your breed.

· Align yourself with a fellow exhibitor or master breeder (someone who has earned a certain number of points at shows and is considered an expert by his or her peers in breeding poultry). If they see real interest in an individual, they are usually more than happy to pass on the secrets of success (everyone wants his or her work to be carried on to the next generation).

BELOW: Judge Paul Kroll of East Aurora, New York, gives a lesson to young exhibitors about showing and judging chickens. After Kroll judged the small number of birds, the kids were to take their turn to see how they compared.

- Learn to be a modest winner and a gracious loser. Don't enter sick, lousy, mite-infested, or injured birds. It will reflect poorly on you and your management decisions.
- Don't complain about the judging. If you're upset about the outcome of a show, go home and gripe about it to your best friend, your parents, or your spouse. They might be able to have insightful, constructive criticism about your next attempt.
- Be practical. Eat your culls, and if you decide instead to make them into pets, pick good ones that are tame and personable. Don't cast off your poor birds on others that have come to you for representatives of the breed.

Raising a Winning Bird

If you've decided to raise your own birds from your own breeding stock, it's best to start with breeders that you've purchased from a reputable source. Barred Plymouth Rocks, for example, have been managed for egg production at the expense of their plumage. If you try to enter a Barred Plymouth Rock that you've purchased from a hatchery that sells laying stock, your bird may not be considered show quality. Consulting the opinion of an experienced exhibitor is the best course of action. They'll be able to explain

LEFT: A bantam White Silkie rooster is ready for judging at the Northeastern Poultry Congress's annual Poultry Show at the Eastern States Exposition Center in West Springfield, Massachusetts.

the correct *type* required for your breed (not just its color, but its correct weight and size, the shape, slope, and carriage of its body and its tail, its feather quality, the shape of its comb and the overall behavioral characteristics associated with the particular breed).

Exhibitors speak of a bird's *bloom*, or its overall health and cleanliness. Obviously, a bird that is old or molting would not be a good candidate for show. There are various trade secrets to bringing a bird to this peak of perfection. Following is a list of some more common conditioning secrets:

- Feed your birds 60% dark leafy greens. Don't allow them to eat so much that they develop loose stools. Don't allow your birds to forage for the greens; sunlight will bleach the feathers, or turn them a reddish hue. Instead, two months before the show, move them to their own pen, inspect them closely for body parasites and broken feathers, and get them used to the pen or coop that is similar to the one they will be exhibited in.

- Animal protein is considered essential for optimum bloom. While breeders and show birds should be fed a maintenance diet of 18 to 20% protein for optimum plumage, some breeders also supplement with other animal proteins. These can be as simple as dry cat food or as specialized as Tizz Whiz Sho Glo, a prepared protein supplement made up of 90% triglycerides and rice bran oil. Cod liver oil is also a good source of animal protein.

- Sunflower seeds, wheat germ, and probiotics (whey milk or yogurt products) are also sometimes used to condition birds.

- Vinegar added to the drinking water is said to change the pH in the chicken's digestive tract and act to attract beneficial microbes for digestion while warding off coccidia.

- If you're raising white varieties or white-wattled breeds like Minorcas, Spanish, or Leghorns, don't feed them corn; it tends to impart a brassy color to their plumage.

How to Wash a Chicken for Show

Believe it or not, most chickens get baths before their shows. Light-colored birds always need washing, while darker, tightly feathered birds may not need dousing unless they're heavily soiled. Most exhibitors like to wash their birds at least three or four days before the show, so that the natural oils have returned and out-of-place feathers can be repaired.

A bathroom or laundry room sink is ideal to wash your bird in; it's not drafty, it's at the right height, and you have access to water that is warm (around 90 to 100 degrees Fahrenheit).

Use only mild shampoo, like baby or hypoallergenic shampoo. Some people prefer to put the shampoo directly into the water for a bubble-bath effect and then put the bird in up to its neck feathers. Others like to put just 3 or 4 inches of water in the basin and wet the feathers down then apply the shampoo directly to the bird. However you decide to do it, use calmness and quietness so that the process does not scare the bird. I like to place my hand under both wings,

BELOW: Cody Schwieger, 8, of Strafford, New Hampshire, has a calming effect on his chicken—a black-breasted red Old English Game—while waiting his turn in the Pee Wee Showmanship Competition.

cupping the breast, gently lowering the bird into the water. When you wash the feathers, don't go against them; work from the head back toward the tail, paying special attention to the vent, legs, and feet. An old toothbrush works great to brush the scales of the legs and feet clean. After shampooing, drain the basin and fill it with warm water again to thoroughly rinse off the bird. Then remove most of the water from the feathers by hand, carefully squeezing them and then towel-drying. A blow-dryer set at a low setting is handy for fluffing up the feathers and keeping the birds from becoming chilled.

You can finish grooming the bird by gently wrapping it in a towel and working on its comb, face, and wattles. A little baby oil, vitamin E oil, or olive oil can be rubbed into the fleshy areas, keeping care not to soil the feathers. A toothpick or cotton swab can be used to clean around the bird's nostrils and ears.

BELOW: Bucket, a Polish Crested pullet, is one of the cherished chickens owned by Brianne Riley and Matthew Taylor, because she survived an overly enthusiastic greeting from one of the couple's dogs. Bucket now visits school classes with Riley.

Preparing Your Bird for Cooping

Any bird that has not been introduced to the confinement of the show pen will be a disaster in the exhibition hall. Not only will the sights and sounds be overwhelming, but the small cage alone is enough to unsettle the bird.

Because chickens love routines and hate surprises, it's a good idea to train your bird for the show. The younger your bird is when you begin this process, the more successful you'll both be.

You should start confining your potential exhibition birds at least two weeks prior to the show. Handling the birds at least two or three times daily will simulate the judge's inspection of your bird. Be sure to remove your bird from the coop by cupping it under its wings near the breast and plucking it from the cage head-first. It should be returned to the coop head-first, so the wings fold back and don't injure their feathers upon reentering. Flighty and nervous birds should be approached quietly and calmly, and if they seem distressed, give them a time-out to get used to the situation.

Every breed has a certain pose it should strike when the judge is examining it. Consulting the *Standard of Perfection* will help you "set up" your bird to strike these poses, and again, the help of an experienced exhibitor is invaluable.

Above all, showing your birds should be a fun event. The prizes won't make anyone rich, but the camaraderie of fellow exhibitors and the commitment to preserving diverse breeds will make it worthwhile.

RIGHT: William Trickett, 13, of Mansfield, Connecticut, eats a yogurt while his brother's Bantam White Cochin pullet sits on his lap.

Bradford Jones, 9, of
Royalton, Vermont, holds
a chick that was hatched
by incubator

(10)

Flock Health: Common Illnesses, Conditions, and Diseases

It's a beautiful fall day on Cape Cod, Massachusetts, 1997. The sky is crystal clear blue, and the air is sharp but still frost-free. Our chicken menagerie has grown, even though my husband and I still live as winter caretakers for six months of the year, in luxury homes vacated by their owners for warmer climes, and then in an unheated barn with no plumbing during the summer months. We have the obnoxious Border Collie who has just chewed my antique hooked rugs to pieces, and the beautiful, white Arabian gelding who loves to canter in the waves. We have our three cats, content with chasing June bugs and Fowler's toads through the swampy lawn that surrounds us in the summer and the catbirds and mockingbirds that flit through the bull thistle and bittersweet in the winter. We sell eggs at the local health food store, and we grow custom chicken meat for a handful of people. And we are still searching for a small farm of our own.

Floyd the rescued white leghorn is happy in her rustic coop, finally able to socialize with the Plymouth Rocks and Rhode Island Reds, but as I look out the window, I spy Pinky, tittering across the lawn. Her head is down, her feathers fluffed up like she's wearing a carnival costume, and she can't seem to stay upright. I rush out to snatch her up, and then bring her inside to consult my chicken health handbooks. It looks for all the world as though she may have something horrible, like bird flu, or encephalomalacia—maybe it's even mycotoxicosis. I can't even open the books for fear of reading the diagnosis, so I just hold her close and try to think.

Pinky is one of my favorites. She's a cross between an Ameraucana and a Leghorn, so she has the stature of the former (full of herself, almost haughty, because she has the ability to lay green-shelled eggs) and the demeanor of the latter (an extrovert, an over-achiever, a multi-tasker who is never quite satisfied with the final product). Only now, she just looks drunk. I cradle her next to me. She smells drunk—her breath actually smells like alcohol, even faintly like orange liquor.

It occurs to me that the compost is very close to the chicken yard, and Pinky is adept in escaping the dog-proof yard, in search of an adventure. I walk to the compost and discover evidence of a late summer's binge: lots of tiny peck marks in odoriferous, fermenting oranges. I plop Pinky in a cage that I label the drunk tank and supply her with lots of drinking water. Sure enough, when I'm brave enough to crack open my chicken books, it warns of giving chickens free access to the compost pile for this very reason.

The next day dawns as clear and as beautiful as the last. Pinky is skittering around the yard in search of grubs and spiders. I putter around the yard, picking up the compost and moving it to a container that is chicken-proof. We decide not to talk of her indiscretion, or of mine.

On the home scale, preventing disease and illness in your flock is much more manageable than on a commercial scale. First, many of the commercial strains of chickens have been genetically selected over the last few decades for maximum production rather than for longevity and overall heartiness. Commercial egg laying facilities have flocks that aren't free-range, are in climate-controlled environments, and are replaced (culled and killed) every 16 months, so not many of the birds succumb to natural causes.

On the other hand, the average age of chickens in the backyard or small farm flock is 4 years old. Natural mortality can appear higher in these flocks, particularly if it's composed of commercial strains of layers or meat birds.

Until you're comfortable caring for your flock, it would be best to choose a breed that is more robust—rather than a maximum production breed—and therefore easier to keep. (Refer to the Chicken Breed Chart, page 38–39, for specific characteristics that would best suit your flock's conditions.)

Additionally, if preventative measures for nutrition and husbandry are followed, most health issues can be avoided. A quick summary of main points mentioned in previous chapters follows:

- Stay away from the "free chickens" offers. You're only asking for problems if someone is trying to rehome their castoffs. Reputable poultry breeders ask a fee for their birds because it's how they make part or all of their living.

- Isolate any new birds coming into your flock. The stress of introducing them to a new location with new birds and a different feeding regimen can cause health conditions that could mimic disease. You want to give the birds time to adjust before thinking that they're truly sick. A period of 7 to 10 days in isolation should be enough transition time before introducing the new birds to the existing flock. Try to introduce them at night, and if you have a free-range flock, keep the whole flock inside for a few days, so the new birds learn where to come back to roost. During this time, give the flock "toys" to distract the old birds from ganging up on the new birds. I give mine over ripe cukes or zucchinis in the summer; in winter, I give them a winter squash to pick at.

- Don't mix birds that are wildly different in age. Chicks can be housed together if they are between 2 and 10 days of age; pullets (hens that haven't started to lay yet) between 10 and 20 weeks can commingle; and introducing the younger pullets to the established flock should occur at night.

- Don't mix high production meat breed chicks and layer chicks. The layer chicks will pick at the meat birds and can even cannibalize them.
- If you lose birds to mortality, especially if the cause of death is unknown, properly dispose of their remains. If hatching your own chicks, either in an incubator or under a broody hen, don't give the eggshells of the hatched chicks to the flock as a source of calcium. Instead, compost them away from the flock (they're great for the soil).
- When you suspect illness or disease amongst your birds, get a professional diagnosis from a reputable breeder or veterinarian. Many medications used to treat illnesses, disease, and even parasites in chickens require withdrawal periods after treatment before the meat or eggs can be safely consumed again.

Common Conditions Affecting Your Flock
BREAST BLISTER

Hybrid broiler chicks available from hatcheries take on such rapid growth rates, particularly in the breast, that their legs are unable to

support their body weight. Many of the commercial meat breeds spend longer periods of time squatting on the ground at the feed trough rather than foraging for food, and their keels become irritated. Just as bed sores can form from constant contact with an irritant, so do breast blisters form. Heavier heritage breeds and poorly feathered breeds are also

LEFT: The bird's crop (where it stores its food before being digested) should be prominent but not bulging, as in this hen's. Chickens with impacted crops often regurgitate undigested food when handled.

susceptible to blistering. Most of the sores can be prevented by providing deep litter that is dry and nonabrasive for birds with rapid growth rates, or by padding nighttime roosts.

CROP IMPACTION

Also referred to as *Pendulous Crop* or *Crop Binding*, this condition is not contagious. It usually occurs in mature or older birds; why it occurs is unknown, although it could possibly be genetic. Injury or improper rations can also cause the crop to lose its muscle tone and eventually become unable to empty its contents. Confined birds that are bored and pick up the bad habit of eating roughage like straw and other bedding can cause their crops to plug and become unable to function. Birds should be provided with fresh drinking water at all times and be provided with proper rations to prevent this condition.

WHAT TO LOOK FOR

This condition is most easily diagnosed from its symptoms: noticeably distended or bulging crops that are filled with sour-smelling or fermented feed and other debris is typical. While crops can become fairly full shortly after a bird has eaten, it should eventually shrink in size. A crop should feel firm and not squishy when palpated. If a bird regurgitates foul-smelling material after the crop has been manipulated, you should suspect an impaction. This is usually a chronic condition, so the bird will overall feel thin and look unthrifty.

FEATHER LOSS

Your birds will naturally lose their feathers, or *molt* once a year. Genetic manipulation has produced laying breeds that wait as late as possible to do this, as when birds are molting, they also cease to lay eggs. Typically, birds will molt late summer, allowing plenty of time for the new feathers to form in preparation for winter. Commercial breeds may not start their molt until late December, so they will need added care to stay healthy, such as higher protein food and a draft-free environment. A molt will generally take 6 weeks

ABOVE: Chickens have an annual molt, where they replace their existing feathers with new ones. Here, the hen is in the "blood feather" state, where they are vulnerable to cannibalism by their coop mates and may have a compromised immune system because so much energy is being put into growing new feathers.

to occur, and it starts with feather loss at the head, progressing to the tail.

Another cause for feather loss is trauma. If a hen is at the bottom of the pecking order, her flockmates may pick at her and cause her to lose feathers. Bored hens can also learn this vice, picking and pulling at another hen's feathers to the point of drawing blood. An entire flock can learn this bad habit, and the blood can spur cannibalism in the flock. Overcrowding, poor nutrition, and overheating can all lead to feather picking.

A rooster can also cause feather loss, particularly if there are too many roosters in the flock with not enough hens. Typically, feathers at the back of the neck and on the back will be missing, as this is where the rooster mounts or "treads" the hen during mating.

FROSTBITE

The comb of a chicken is highly vascular (filled with blood), and very susceptible to freezing, much like our fingers and toes are if not protected. Roosters and large-combed breeds like leghorns can't tuck their entire comb under their wings when sleeping, so they can suffer during the cold winter months. Chickens are capable of tolerating a fair amount of cold, especially the heavier breeds like Australorps, Orpingtons, and Plymouth Rocks, but damp, drafty conditions combined with cold or frozen drinking water can cause frostbite. The stress and pain associated with frozen combs can cause lower production and general unthriftiness. Chickens may become depressed and uninterested in eating and drinking. Coop mates may also peck at the individual, so badly affected birds should be isolated from the flock. I would not recommend drastically changing the environmental conditions unless the bird is seriously ill. The stress of returning to the coop after healing may be just as detrimental to a bird that has acclimated to winter living conditions. Many professional poultry owners will cut the comb off, called *dubbing*, to avoid the chance of frostbite.

WHAT TO LOOK FOR

Both the comb and wattles are susceptible to freezing; they will at first become swollen and look pale in color. Eventually, parts of the comb or wattles will blacken and then shrivel and fall off. Applying an over-the-counter topical antibiotic on the frozen areas will prevent infection, though offering foods rich in vitamins and minerals as supplements is usually sufficient medicine to pull mildly frostbitten individuals through the worst of it.

PICKOUT/PROLAPSES

Also called *blowout* or *prolapsed oviduct*, this condition occurs when the lower part of the hen's reproductive tract turns inside out and protrudes from the vent (the bottom). It can happen when hens that are too young are forced into laying, when the birds are too fat, or if they are laying larger than normal eggs because of some other reproductive disorder. Treatment using Preparation H or any other hemorrhoid cream applied to the area will usually resolve the problem if it's caught early enough and if the hen is isolated from other chickens so that she's not picked at.

WHAT TO LOOK FOR

An abnormal protrusion from the vent, typically bright red and bulbous. The hen may also have a soiled vent from stool being forced out around the prolapse.

VENT GLEAT

While the definitive cause for this is unknown, it is a condition that is quite common in chicks less than 10 days that are not reared by a broody hen, and sometimes seen in older laying hens. Both improper brooding temperatures, which results in chilled chicks, and inconsistent rations are often associated with vent gleat.

WHAT TO LOOK FOR

Droppings become "pasted" around the vent and can cause it to seal shut, resulting in the chick's death. If chicks are found with vent gleat, carefully pull off the down from around the vent, and express the fecal material from the rectum by gently pushing below it, near the chick's belly. Carefully inspect all of your chicks daily, as they are very rarely all affected by vent gleat at the same time.

Common Diseases

There are volumes written on poultry diseases, mainly for the commercial industry, because of its economic importance. Many diseases have similar symptoms, so diagnosing it successfully without a professional lab is almost impossible.

What should be kept in mind if you suspect disease are the following: is the problem isolated to an individual, or are you seeing coop-wide problems? One sick bird will most likely point to a less serious problem than if the whole flock is affected.

AVIAN INFLUENZA OR BIRD FLU

The virus that causes this disease has many strains; most infectious is strain H5N1, and it is of particular concern because of its highly contagious nature and the fear that it can mutate and be passed from bird to human (called a *zoonotic* disease). The media has often used this apocalyptic fear to stir up a frenzy, causing many people around the world to kill or sell their poultry, despite their dependence on the flock for income or sustenance. Often targeted as potential harbingers of this disease are backyard flocks, the logic being that the small flock owner will not recognize the signs of the disease. This makes no sense to me, as I believe that small, diversified populations that are well cared for are far less likely to contract disease than overcrowded, monocultured poultry houses, filled with all of the same genetics and susceptibilities.

WHAT TO LOOK FOR

Much like the virus that causes the "flu" in humans, avian influenza is highly contagious in birds. It is contracted when an uninfected bird comes in contact with the feces or expelled air from infected birds. The only way to avoid avian influenza in your flock is to prevent contact with infected poultry or infected wild birds; there is currently no strain of H5N1 in the United States.

Unfortunately characteristics of the infection mimic many other diseases: respiratory problems, unthrift, depression, and a lack of appetite. Diarrhea, swelling around the face, and reddish or white spots can also be associated with the disease. Only a specialized poultry laboratory can diagnose your flock with this disease. Certainly, if all of your birds are ill, it would be in your best interest to test one of your sick birds for this disease.

BUMBLEFOOT

This disease occurs when a bacterium naturally occurring in chickens, *Staphylococcus*, *Escherichia coli*, or *Streptococcus* enters a portion of the chicken's foot, creating an infection.

RIGHT: Chickens have small air sacs on either side of their beak and below their eyes. When they become infected, they swell.

Because bird's bones are hollow, the progression of the disease can be debilitating. An injury, improper bedding, rough roosts, rocky poultry yards, wire flooring, and even genetic disposition can all be culprits that lead to the infection. Once it has started, a pus-filled, black area in the foot pad forms, and sometimes even the hock joint will swell. The disease is very difficult to cure, so prevention is the best management: using deep layers of dry, soft litter, and choosing hearty breeds will avoid problems with bumblefoot. Washing the affected area and using a topical antibiotic can stop the spread of disease if caught early.

CHRONIC RESPIRATORY DISEASE AND INFECTIOUS SINUSITIS

These diseases are both spread by the same organism, *Mycoplasma gallisepticum*, which is present in all parts of the United States and can affect many different species of poultry. The disease is spread through the eggs of infected carrier hens, but most commercial flocks are now free of the disease. Although high mortality is not usually associated with these diseases, it can lead to reduced growth rate and a lack of feed efficiency in meat breeds and a reduction in the overall production in laying birds. Antibiotics can be used to successfully treat the disease, but birds that recover should not be used for breeding or replacements.

WHAT TO LOOK FOR

With both diseases, coughing, rattling sneezing, and difficulty breathing may be seen. There may be discharge from the eyes that is yellowish or foamy. The area below the eye and above the beak may be swollen or discolored. Birds may have a lack of appetite or have a reduction in productivity, both in growth and in egg laying.

DISEASES ASSOCIATED WITH CONTAMINATED OR POOR QUALITY FEED

Chickens are especially sensitive to toxins associated with their feed. In the wild, birds eat fresh berries, plants, live insects, snails, even baby mice, snakes, and other unprocessed foods. When they are presented with a formulated feed, problems can arise. (Refer to Chapter 4 for more on this subject.) *Mycotoxicosis* is a disease caused by consuming feed containing toxic byproducts of molds. Many molds are grain-specific, and some prefer higher temperatures in conjunction with moisture. It can be hard to pin the problem on your grain, because after it leaves the mill, the storage conditions can contribute to conditions that favor the growth of mold.

One way to avoid *mycotoxicosis* is to store grain in containers that don't attract moisture or high heat and favor low humidity. Another is to purchase commercially prepared feeds with mold inhibitors. Or, if using organic grains, only purchase an amount that you will be sure to use within a month. Symptoms of *mycotoxicosis* mimic several other diseases, so analysis of your feed would be necessary to rule out other possibilities. Of course, if your grain smells or looks moldy, you should suspect your feed as the culprit.

The organism *Clostridium botulinum* is responsible for botulism. Although it is more commonly found in waterfowl, other poultry can become affected. The organism grows on spoiled or decaying feedstuffs and produces toxins that affect the nervous system, specifically around the neck, which is why this disease is sometimes referred to as *limber neck*, because the affected birds lose control of being able to properly position their necks.

Crazy Chick Disease, also called *Encephalomalacia*, is a nutritional disease that most commonly affects young birds between 2 and 4 weeks of age. Although the disease is notably rare in commercially fed flocks, it can be more of a problem if birds are fed organic rations, which don't contain artificial preservatives.

Natural antioxidants like vitamin C and E can become depleted quickly in organic feed if they are used up trying to stabilize other components in the rations that are susceptible to becoming rancid in high heat or humidity, like soybeans. A deficiency in vitamin E often goes along with a selenium deficiency, both of which can cause a young bird's brain to swell and deteriorate. If caught early enough, the affected birds can be treated with a vitamin supplement in the drinking water, or with an injection of vitamins A, D, and E under the skin. The disease is easily prevented by using only fresh rations that are stored in cool, dry locations (where oxidization is greatly reduced). Using a vitamin supplement in the drinking water when hot and humid conditions persist will also guard against an outbreak of this disease.

WHAT TO LOOK FOR

This is a rapid-onset disease, where chicks appear normal one day and sick the next. The chicks will appear "drunken," and stumbling, unable to stand, may flap their wings uncontrollably, bend their heads back, and wag them back and forth. The condition eventually leads to paralysis and death, usually within a matter of 12–24 hours.

MAREK'S DISEASE

This disease is very common in large-breed chickens, though Sebrights and Silkies can also become affected; flocks that are unvaccinated are particularly at risk for the disease. Most poultry houses vaccinate their day-old chicks because of the severity of the disease and its associated high mortality (as high as 80 percent). Spread through feather dander, it is a strain of chicken herpes called *Gallid herpesvirus 2*. The virus is capable of traveling miles through the air, and once it is in your poultry yard it can persist for up to a year, with carrier birds that can shed the virus with no signs of the disease. The disease can cause tumors in the skin, muscle, and organs such as the spleen, liver, lungs, kidneys, and heart (these would only be visible during a necropsy).

WHAT TO LOOK FOR

Because Marek's can involve the nervous system as well as muscle, birds that are over 2 weeks of age can show signs of paralysis, stunted growth, abnormally red-colored feather shafts, pale skin, or a stilted gait. Changes in the iris can also occur; instead of a vibrant reddish-orange color, the iris turns grayish and mottled.

A blood test can indicate whether your birds are resistant to the disease; vaccination is the only prevention.

MUSHY CHICK DISEASE/ OMPHALITIS

Also referred to as *Navel ill*, or *Navel Infection*, this disease is caused by the bacterium, *E. coli*, which causes a condition called *Colibacillosis*.

WHAT TO LOOK FOR

In unhatched eggs, fully formed chicks will be dead in the shell. Newly hatched chicks will fell wet or "squishy," even after they have dried from hatching. Older chicks can be affected, and their navels will look unhealed, red, swollen, or distended.

There is no cure for the disease once contracted; hatching clean, uncracked eggs in an incubator with proper humidity and temperature prevents this disease. Keeping newly hatched chicks clean, dry, and at the proper brooding temperature is also necessary.

PULLORUM

Caused by the microorganism *Salmonella pullorum*, this disease affects young chicks resulting in extremely high rates of mortality in birds under 2 weeks of age. It is sometimes referred to as *White Diarrhea*, although several other conditions can cause the feces to appear white. It is passed from the diseased ovaries of a hen to her egg, and the chick is hatched already infected; chicks can spread the disease in an incubator or brooder through their down. While the hen has the disease confined to just her ovaries, nearly all of a chick's organs can become infected. Pullorum occurs worldwide, and can affect other species of poultry. There are many blood tests available to test adults for the disease, and sulfa drugs can reduce mortality in the flock. However, treating your flock without diagnosing the disease can establish carriers. Birds that test positive for Pullorum should not be used as breeding stock (it is for this reason that you should always test your birds for this disease if you intend to use them as replacements or for breeding).

WHAT TO LOOK FOR

Infected chicks show symptoms of Pullorum shortly after hatching. They almost always have a whitish diarrhea pasted around their vents, and they stand droopily with their eyes closed and feathers ruffled. On the other hand, birds that recover from the disease may show no symptoms at all, but a post-mortem of an infected bird reveals the ovaries to be dark, shriveled, and shrunken. In the chick, the liver looks yellow and soft rather than firm and dark. In some cases, the sac around the heart can be distended with fluid.

Parasites

Chickens are naturally infected with both internal (living inside the bird's body) and external (living outside the bird's body on the feathers or skin) pests, particularly if they are in a free-range or backyard coop environment. Wild birds, earthworms, blades of grass, other animals, and even the soil where chickens forage can contain parasites. An infestation can cause grave health problems, but worrying about keeping your flock completely parasite-free isn't necessary. Usually, chickens that have been provided with good sanitation and proper nutrition with a diet high in vitamins and minerals can build immunities to ward off infestations or overloads of parasites. Parasites depend on their hosts to continue surviving,

so setting up a balance that benefits their survival and the chicken's is in their best interest. Sick, older, or broody chickens, and birds exposed to stress can all lead to conditions that can cause parasite outbreaks that birds cannot overcome.

If you suspect an infestation, consult your local grain supplier or veterinarian for appropriate treatment: don't assume that because your bird looks down and out it has worms. Remember that proper husbandry will help you

LEFT: Soiling around the vent, or cloaca, can indicate parasites or other infection.

avoid the use of chemicals or medications that are used to control parasites. Keep your chickens clean and dry, especially younger or older birds, and maintain them on a healthy, free-ranging diet.

Following is a list of common internal and external parasites. Again, if you suspect a problem, consult a professional before treating your birds.

COCCIDIOSIS

Also referred to as *Coxy*, this is a very common affliction caused by a single-celled parasite called a protozoan (amoebas are protozoans). It can be responsible for mortality in a flock that ranges from mild to severe, particularly in younger birds that are between 3 and 6 weeks of age. Coccidia naturally occur in the environment and in mature birds. There are nine different species, and of these, three are the most detrimental to the health of your flock. These parasites affect the digestive tracts of the birds, hence the severity of infestation in young birds whose GI tracts are still developing. Coccidia naturally occur in the soil and in the litter (shavings, sawdust or other substrate) that you house your birds on. For a completely coccidia-free environment, your birds would need to be reared in a wire cage, away from the soil and natural environment. Instead, proper husbandry and vigilance can keep your flock from an outbreak of coccidiosis.

Many hatcheries offer vaccines for coccidia, but just as the flu shot for humans is developed for the most commonly occurring strains of the virus during that season, the coccidia vaccine is not 100% effective. There are also many feed rations that offer coccidiostats; if you keep your birds clean and dry, you won't have problems and will end up relying on far fewer stop-gaps to raise your birds.

WHAT TO LOOK FOR:

Birds that are infected with coccidia will be fluffed up, feathers ruffled and rough looking, listless, and show no interest in eating or drinking. Their vents will be soiled, and the feces may be flecked with blood. Taking a sample of the feces to your veterinarian can diagnose the presence of coccidia (the protozoans are only visible with a microscope). Coccidia thrive in a hot, humid environment, so taking care to provide fresh, clean water, a dry pen and prevention from overcrowding can avoid an outbreak. Eventually, chicks build up immunity to this parasite. Some flock keepers put apple cider vinegar in the drinking water to fortify the intestines (1 tablespoon per gallon of water), as they claim it changes the pH in the gut and reduces the ability for the parasites to affect the absorption of nutrients.

If you diagnose a coccidia outbreak, isolate the diseased birds, as their feces will continue to shed parasites and potentially affect the rest of the flock.

ABOVE: Having a variety of species that pasture together can aid in the prevention of parasite infestation and predator threats. Here, chickens and pigs cavort at Fat Rooster Farm.

WORMS

Horror stories of all kinds have been associated with this five-letter word. Tales of cracking open an egg and finding a long, spaghetti-like strand send fear into the hearts of many small flock owners, when really it is incredibly rare and would occur only if the flock were severely under-cared for.

That aside, poultry are susceptible to infestation by a large number of intestinal parasites whose effects can result in loss of condition, productivity, an increased vulnerability to other disease, and, in some cases, death.

There are two principal types of worms that affect chickens: roundworms and tapeworms. Roundworms (in the Nematode family) are round and long and have no segments. They are usually gray or white in color. Tapeworms (in the Cestode family) are long, flat, and segmented. Mature tapeworms live in the gut, but shed "gravid proglottids," the little broken off segments, which are little packets containing hundreds of mature eggs. If you see worms in your flock's feces, the general rule of thumb is that if it looks like spaghetti, it's in the round worm family; if it looks like rice, it's in the tapeworm family. For a complete diagnosis, you should take a sample of the feces and the worm to your veterinarian; not all medications that control and kill Nematodes will kill Cestodes.

WHAT TO LOOK FOR:

Younger birds are most severely affected by worms; their gastrointestinal tracts are still developing and their immunities are not as effective in preventing infestations from occurring.

Unless you have a heavy infestation, you won't necessarily see mature worms in the feces. Take a sample of the stool to your veterinarian, so they can look for eggs under the microscope and provide you with the correct medication. Other signs include stunted growth, emaciation, and general unthriftiness.

One type of worm, *Syngamus trachea*, or gapeworm, attaches itself to the trachea (air pipe) of the chicken. An infestation causes the bird to sneeze, stretch its neck, and "gape" its mouth open because of the inflammation and mucous caused by the worms.

Another worm, *Heterakis gallinae*, or cecal worm is associated with the disease **Blackhead** or **Histomoniasis,** which can be fatal in turkeys cohabitating with the flock. The mature worms live in the cecal folds (GI tract) of chickens, ducks and geese. A parasite called *Histomonas meleagridis* invades the cecal worm and then is passed to turkeys, where the parasites migrate to the liver and intestines, causing death. Because both parasites are needed to cause the disease, if you have a closed flock and don't import chickens from areas that may have this worm, safely keeping turkeys in your flock shouldn't be a problem.

ABOVE: Turkeys can happily co-exist with chickens, providing blackhead disease has not been a problem on the farm. Chickens are carriers and can infect turkeys.

For all worm infestations, the key to prevention is clean water and food, and dry, clean litter.

LICE AND MITES

Several different species of lice can be found in chickens, depending on your geographic location. Lice in chickens chew and bite, but do not actually suck blood from the bird. Still, they can be irritating and cause an unthrifty appearance. Chickens naturally preen their feathers several times during the day, but older, sick or broody birds don't and can therefore become infested. Lice are host-specific, so if you handle a chicken that has lice, they may crawl on you, but they won't take up residence. Our body temperatures are far too cold to sustain them for very long. A louse will stay on the bird for its entire life, and it is visible on the skin and feathers. I inspect a bird for lice near the eyes, the mantle feathers (behind the head), and near the vent. Lice are fast, so you need to look quickly. They lay their eggs typically near the vent or the wattles. They look like columns of grayish white eggs huddled in clusters.

Chickens naturally prevent infestations of lice by dust bathing. You can provide your birds a dust bath of soil if they don't have access to it. Many flock owners also give their birds *diatomaceous earth* to bathe in. Diatomaceous earth consists of fossilized remains of diatoms, a type of hard-shelled algae. It is naturally abrasive, so the theory is that it will penetrate and damage the body of the louse while the bird is dust bathing in it. However, prolonged inhalation of the siliceous remains of these diatoms can cause lung damage in humans, so it should be used with caution.

Wild birds are naturally infected with lice, so if you are free-ranging your birds they are bound to have lice. However, young, sick, older, or broody birds can be adversely affected by lice. If you feel that your flock's health is being compromised, there are several louse medications that will successfully treat an infestation. Again, consult a professional before treating your flock.

ABOVE: Scaly leg mite on a mature bird versus normal leg conformation.

Mites are bloodsuckers and can cause weight loss, anemia, and loss of egg production. Some mites are only active at night. They are smaller than lice, typically brownish or red in color and may live in nest boxes, cracks in the walls of the coop, or near the vent, tail, or breast of the birds themselves.

The *Scaly Leg Mite* (*Cnemidocoptes mutans*) is an exception. It generally confines itself to the legs of poultry, burrowing under the skin, and giving the scales a rough, grayish appearance. If a severe infestation occurs, it can cause weakness, reduced production and loss of appetite. A tried and true cure for this malady is to treat the infected bird by dipping its legs into a mixture of 1 part kerosene and 2 parts oil, such as mineral, raw linseed, or petroleum jelly. By treating once or twice weekly until the problem is resolved, the mites are suffocated and eventually die. Ivermectin, a parasiticide available at your veterinarian, also kills the mites.

ABOVE: A healthy, barred rock cross struts his stuff at Fat Rooster Farm in Royalton, Vermont.

Victoria Rickard, 7, of Hampton Bays, New York, reacts to placing second with "Snowflake," her White D'Uccle in the Pee Wee Showmanship Competition. Rickard placed first in the year before, her first time showing chickens.

Glossary

American Standard of Perfection—The publication by the American Poultry Association that describes in detail each of the organization's recognized breeds.

Avian—Anything having to do with birds.

Bantam—Most Bantams are miniature chickens that have been bred to be about a quarter of the size of the large breed with the same name. In some cases, such as the Rose Comb and Silkie, Bantams don't have large-breed counterparts.

Banty—Nickname for Bantam chicken.

Barnyard Classic—A chicken hatched on the farm of mixed parentage (a mutt).

Biddy—An old-timer's nickname for a hen, usually one that goes broody often.

Bleaching—A predictable sequence of fading yellow in laying hens that can be used to predict how long they have been actively laying.

Bloom—The chalk-like, protective coating on an egg which quickly dries after laying and acts to seal out bacteria and dirt. Also called the **cuticle**.

Breeds—A group of chickens with shared characteristics, such as comb type and plumage, that differ from other groups of chickens.

Broiler/Fryer—A meat bird, typically a Cornish Rock crossbreed, that has been slaughtered at five to ten weeks of age and has a weight of 3 to 5 pounds.

Brood—To take care of a batch of chicks; or the hen's babies (i.e., her brood).

Broody Hen—A hen that has laid a clutch of eggs and has begun sitting on them (setting) with the intention of hatching. Once hens become broody, they cease to lay eggs and will not resume laying until their chicks have reached independence (the age of independence varies according to breed). Broody hens will readily accept eggs laid by other hens.

Candle—To check the contents of a whole egg using a light source.

Cannibalism—A bad habit, typically brought about by boredom, overcrowding, or disease, where other chicken's eggs, feathers, or flesh are picked at. The persistent pecking causes wounds that eventually become infected and may lead to the chicken's death.

Capon—A surgically castrated rooster. This practice is now less common with the improvement of growth and product quality in the meat chicken industry. Historically, it would allow the home meat raiser to keep the birds for a longer period of time without developing secondary sex characteristics, such as crowing, aggression, or a tougher flesh. Capons are still available in grocery stores, typically weighing 15 to 18 pounds dressed at less than 25 weeks of age.

Chalazae—The two cords that suspend or anchor the yolk in proper position within the egg white (singular is chalaza).

Chicken—A member of the poultry family. The term can be used both to refer to a rooster or a hen, as a horse can refer to either a stallion or a mare.

Chicks—Baby chickens. Chickens, like all birds, are hatched, not born.

Cloaca—Latin for *sewer*, it is the passageway just inside the vent where urinary, fecal, and reproductive tracts merge.

Clutch—A group of eggs laid by a setting (broody) hen, usually numbering between 6 and 15.

Coccidiosis—A protozoal infestation causing disease and death, typically in young chicks, and usually associated with unclean housing conditions. High heat and humidity exacerbate the probability of an infestation occurring.

Coccidiostat—A drug that is used to control coccidiosis.

Cock—A male chicken, also called a *rooster*.

Cockerel—A young, male chicken, under one year of age.

Cornish Game Hen/Rock Cornish Hen—A meat bird, typically a Cornish Rock crossbreed that has been slaughtered at two to three weeks of age. Chicken products labeled as Cornish Game hens are not necessarily females.

Crop—The esophageal pouch that lies below a bird's neck, where food is stored directly after being eaten and before it travels to the rest of the digestive tract. Also refers to the act of trimming the bird's wattles (see *Dub*).

Cull—A hen or rooster that is no longer needed in the flock and is therefore removed and either slaughtered for consumption, or composted. Removal is based on age, productivity, health, or disposition. The act of culling refers to removal of unwanted chickens from the flock.

Cuticle—see *Bloom*.

Debeak—To remove a portion of the top of the beak to prevent injury when a bird attempts to cannibalize its coop-mate.

Droppings—Chicken poop.

Dual-Purpose Breed—A heritage (old-fashioned) breed that is useful for producing both meat and eggs but does not necessarily excel in the production of either when compared to breeds developed specifically for eggs or meat.

Dub—To trim the comb, typically of a rooster, to adhere to show standards, much like ears or tails are cropped in some show dog breeds.

Dust—The activity chickens use to clean their feathers and rid themselves from body parasites by "bathing" in soft, loose soil.

Egg Tooth—The hard, temporary cap on the upper part of a chick's beak that it uses to crack its way out of the egg. The tooth falls off within the first week of hatching.

Feces—see *Droppings*.

Fertilized—An egg that is capable of producing a chick.

Flock—A group of chickens living together.

Fryer—A young meat chicken, also called a *broiler*.

Gizzard—The organ that contains grit and is responsible for grinding up the hard foods that chickens eat.

Grade—The act of sorting a hen's eggs, based on shape and size.

Hatch—The act of emerging from the shell, or the event of a group of chicks in a clutch coming out of their shells at the same time.

Hatchability—The percentage of eggs that hatch if incubated, depending on such factors like condition of the breeding stock, shape and size of the egg, and seasonality of lay.

Hatching Egg—A fertilized egg that has been properly stored for incubation so that its hatchability has not been compromised.

Hen—A female chicken, over one year of age, capable of producing eggs.

Hen Feathered—A male chicken that has feathers that are hen-shaped (rounded) rather than pointed like a rooster's typically are.

Hybrid—Chicks hatched from a hen and a rooster of different breeds.

Incubation—The act of sitting upon eggs with the intention of hatching. The average incubation period for chickens is 21 days (Bantams tend to hatch a day or two earlier). By comparison, the incubation period for ducks is 28 days.

Laying Hen/ Layer—A breed that has been selected or developed for its ability to lay a large number of eggs annually.

Litter—Bedding used on the chicken house floor and in nesting boxes to absorb moisture, such as pine shavings, shredded newspaper, or straw.

Meat Bird/Meat Chickens—A breed that has been selected or developed to gain weight and produce meat in a short and efficient period of time.

Nest Egg/ Fake Egg—An object that looks like an egg that is put in a nesting place to entice a hen to lay her egg there.

Pecking Order—The social hierarchy of individual chickens within a flock.

Pinfeathers—New feathers that have just emerged from the skin and may still be encased in the waxy shaft so that only the tips are visible.

Pip—The hole in the egg the chick makes with its egg tooth as it begins its hatch; also the act of making the hole (pipping).

Poultry—Chickens, ducks, geese, turkeys, quail, pheasants, and other domesticated birds that are raised for meat, feathers, eggs, or show.

Pullet—A young, female chicken, under one year of age.

Purebred—Chicks hatched from a rooster and a hen of the same breed.

Roaster—A bird that weighs more than 5 pounds and is slaughtered before 20 weeks of age.

Rooster—A male chicken, older than one year, capable of in-seminating (mating) a hen and fertilizing an egg. Males are larger than the female within the same breed.

Set—To keep eggs at a certain temperature so that they will hatch (synonymous with *brood*).

Setting—The act of incubating eggs by a hen, or a group of eggs to be hatched in an incubator or by a hen.

Sexed—Newly hatched chicks whose sex has been determined and are divided into pullets and cockerels. Hatcheries are typically 95 to 99% correct in sexing chicks. Sex of chicks cannot be determined prior to hatching; also called cloacal or vent sexing.

Sex-Linked—An external trait used to determine the sex of newly hatched chicks, such as color or feather characteristic. Barred Plymouth Rock hens crossed with Rhode Island Red roosters will produce chicks with white spots on their heads if they are cockerels; silver females (like a light Sussex) crossed with gold males (such as Rhode Island Reds) will produce chicks that are gold if pullets and silver if cockerels. In the broiler industry, chicks with well-developed flight feathers at the time of hatching are typically females and can be separated from the faster-growing males.

Star-crack—The stage of hatching that occurs just before pipping, and resembles a star-shaped crack in the egg's shell.

Straight Run—Newly hatched chicks that have not had the sex determined; also called as hatched or unsexed. These are usually less expensive than sexed chicks and typically 50% cockerels and 50% pullets.

Trio—One cock or cockerel grouped with two hens or pullets of the same breed and variety.

Epilogue

THE CAT MAY HAVE BEEN SHOT LAST NIGHT.
It looks like a raccoon, which is what my husband was after, down where the meat chickens are housed. He found two dead chicks, bloody and tattered, and a big raccoon hurried away from the bodies, off toward the barn. He tracked its ghostly, yellow eyes to the manure pit, where he let off a blast with the gun, into the cattails, into the darkness. But the cat has not returned, and despite numerous searches, we can't find the raccoon's body, or the cat.

On top of that, there are the skunks. Last year, it was rats and woodchucks. They devastated the dry beans, chewed on the melons, ransacked the grain bins in the barn, even ate the sweet peppers. We were never able to clearly assign blame to one creature or the other, but I'm pretty sure the destruction was a convivial affair.

This year it was different. The skunks started with picking the ripest ears of sweet corn. They deftly stripped the stalks, so as not to disturb the whole plant and its secondary ear which would ripen in a couple of weeks. Each skunk-eaten cob was then shucked as though prepared by a chef for a Maine coastal clam fry, the late-July type, where lobsters are soft-shelled, but succulent, and sweet corn is part of the package. Even an eight-year-old boy can't glean the corn's kernels as cleanly as a skunk.

We trapped four of them in the Ruby Queen corn: a mother and three babies. The Ruby Queen is a special variety of corn that has kernels the color of garnets against green, so I suspect that the skunks were bejeweled with magenta corn milk during their raid, like lipstick on a shirt collar, a taunting reminder of what has been before, and what could be again if the right conditions were to unfold.

Once inside the trap, each skunk made a nest of anything that it could pull in from outside: electric fence, leaves, grass, sod. Each was driven miles away to a new location, on account of the yellow-belliedness of us two farmers, who have an easier time of killing animals that we intend to eat than just outright killing animals because they have a love of corn. The skunks probably wreaked equal havoc in their new surrounds.

LEFT: David Russell, 8, of Orwell, Vermont, holds Juan Carlos, an Old English Game Bantam rooster the author brought for a workshop given to children of parents attending the NOFA-VT Annual Conference in Randolph Center, Vermont.

My neighbor, who claims that even a snail can walk a mile, thinks they just came back home; of course, the practical side of killing a skunk in a trap also has its drawbacks.

By late summer, when the green of the season had turned dark and tired, and animals and people alike begin to prepare for the long, cold winter ahead, the skunks turned to killing the chickens. For this, I have no tolerance, and I began a midnight to 3 AM vigil on the grounds, complete with the 20-gauge shotgun that my husband used to hunt pheasant in southern Ohio.

On a Tuesday, I heard the meat bird chicks begin to protest, in the dark mist of early morning. They're in a remote corner of the farm, past the back part of the barn, on clover-covered pasture. They have an old truck cap to sleep in, but they're not exactly tucked away at Fort Knox. And the corn that the skunks were feasting on there had long been harvested, leaving only yellow nasturtium blooms and the orange glow of sugar pumpkins as options. The chicks and their grain were much more enticing, it seems.

I am wearing one of the long-sleeved camisoles, cast-offs from my clothes-hound city sister. This one is smooth as silk and has a pocket near the left breast. Why these fancy negligees would have pockets puzzles me. What is the pocket normally used for?

For me, the pocket is just perfect. For two backup shotgun shells. In the left hand is the gun; in the right is the flashlight. The gun is loaded with one shell, but just in case, that pocket has the ammunition necessary to finish the job.

And so off I go, out to the meat bird pasture. It's 2 AM, and I've just heard them squawk. The gun is parked near the door with its trigger lock; the ammo is in my underwear drawer upstairs, so our son has no way of connecting the two . . . yet. As I leave the bed, my husband sighs a long, slow, "Oh boy, here she goes again," but honestly, I have no use for these things killing my chickens.

I believe in accountability; a sustainable economy has to factor in personal accountability. No "Oh-my-coffee-spilled-and-burned-me" litigation, no government to blame for not keeping us safe: each and every one of us accountable for our own actions and decisions. Raising meat chickens means protecting the meat chickens, and not blaming the *vagaries of weather, skunks, and pestilence for their failure are all an important part of being personally accountable to me.*

And these are the thoughts I have as I trudge into the pitch-dark night, in my negligee toward the pen. I arrive, and there is the skunk, bent over a mangled chick, not paying much attention to me. Then, when he finally senses me, he runs off, and I lose him. I work the flashlight over the clover field, looking for a black-and-white shape, but only the terrified faces of meat bird chicks greet me. I don't think I ever knew that they blinked.

They are blinking now and slowly begin peeping a high-pitched question to me. "Am I okay? I'm okay? Am I okay?" It sounds more like "Bee be beep? Bee bee bee beep? Bee bee beep?" But that's really what it means.

I can't find the skunk. So I return to the bloody chick's body, ready to pick it up and pitch it into the compost, when it blinks at me. I reach down to touch it. It panics and reels into the dark, one leg dragging, and blood cascading from its back. I find it again, and crouch down. Then I talk. A human, talking to this tiny, battered bird, and it listens. I tell it that I'm going to pick it up and take it to safety, and it responds, small, low sounds of cooing, hoping to get warm. I scoop it up and cradle it in the bend of my right arm, the gun and flashlight held in the other. I have bridged the gap between predator and prey, farmer and food, with just the tone of my voice.

Three nights later, after the cat has gone missing, my son and I sit, perched on the roof of the barn, peering into the cattail-covered manure pit. He has his flashlight, searching for the yellow reflection of eyes, searching for the cat. Perhaps, he says, if we were high up in the air, we could see his eyes glowing. He could tell us where he is. We could take a picture, and there he would be, looking at us. I look up to see the beginning points of yellow starlight reaching through the gray of dusk descending. They look like hundreds and hundreds of blinking eyes, watching us, showing us where they are, telling us where we are.

Appendix 1

Bibliography

American Poultry Association,
American Standard of Perfection.
Mendon, MA:
Global Interprint, 1998.

Belanger, Jerome D.,
*The Homesteader's Handbook to Raising
Small Livestock.* Emmaus, PA:
Rodale Press, 1974.

Claiborne, Craig,
The New York Times Cookbook.
New York, NY:
Harper and Row Publishers, 1961.

Damerow, Gail,
The Chicken Health Handbook.
Pownal, VT: Storey
Communications, Inc., 1994.

Damerow, Gail,
A *Guide to Raising Chickens.*
Pownal, VT: Storey Communications,
Inc., 1995.

Florea, J. H.,
ABC of Poultry Raising. Mineola, NY:
Dover Publications, Inc., 1977.

Graham, Chris,
Choosing and Keeping Chickens.
Neptune City, NJ:
T. F. H. Publications, Inc., 2006.

Hupping, Carol, and the
Staff of the Rodale
Food Center,
Stocking Up.
New York, NY:
Simon and Schuster, 1986.

Klein, G. T,
Starting Right with Poultry.
Charlotte, VT:
Garden Way Publishing, 1947.

Lee, Andy,
*Chicken Tractor, the Gardner's Guide
to Happy Hens and Healthy Soil.*
Shelburne, VT:
Good Earth Publications, 1994.

Lippincott,
William Adams,
and Leslie E. Card,
Poultry Production.
Philadelphia, PA:
Lea & Febiger, 1935.

Luttman, Rick and Gail,
Chickens in Your Backyard.
Emmaus, PA:
Rodale Press, Inc., 1976.

Percy, Pam,
*The Field Guide to
Chickens.* St. Paul, MN:
Voyageur Press, 2006.

LEFT: As Rick Schluntz and Beezer the dog walk into the house, Whitey the White Plymouth Rock hen wanders the family's 1/3-acre in-town yard. While Whitey is older and has slowed down in laying eggs, she's with the family for the duration of her life. "They live with us until they die," said Carol Steingress, Schluntz's wife.

Powell, Edwin C.,
Making Poultry Pay.
New York, NY:
Orange Judd and Co., 1904.

Raymond, Francine,
The Big Book of Garden Hens.
Suffolk, UK:
Kitchen Garden Books, 2001.

Saunder, Simon M.,
Domestic Poultry: Being a Practical
Treatise on the Preferable Breeds of
Farm-yard Poultry. New York, NY:
Orange Judd and Co., 1868.

Stromberg, Loyl,
Poultry of the World.
Port Perry, Ontario,
Canada:
Silvio Mattacchione & Co., 1996.

ABOVE: Arabian mare Molly Rocket grazes among the chickens in the barnyard in South Royalton, Vermont.

Appendix 2

Poultry Organizations and Clubs

Alabama Poultry and Egg Association
P.O. Box 240
Montgomery, AL 36101
334-265-2732
www.alabamapoultry.org

Alberta Turkey Producers
212 8711A-50 Street
Edmonton, Alberta T6B 1E7
Canada
780-465-5755
www.albertaturkey.com

American Association of Avian Pathologists
382 West Street Road
Kennett Square, PA 19348
610-444-4282
www.aaap.info

American Association of Meat Processors
P.O. Box 269
Elizabethtown, PA 17022
717-367-1168
www.aamp.com

American Farm Bureau Federation
225 Touhy Avenue
Park Ridge, IL 60068
847-685-8600
www.fb.org

American Society of Agricultural and Biological Engineers
2950 Niles Road
St. Joseph, MI 49085-9659
616-429-0300
www.asabe.org

American Veterinary Medical Association
1931 N. Meacham Road, Suite 10
Schaumburg, IL 60173
847-925-8070
www.avma.org

Animal Health Institute
1325 G Street NW, Suite 700
Washington, D.C. 20005
202-637-2440
www.ahi.org

Arkansas Poultry Federation
321 South Victory Street
Little Rock, AR 72201
501-375-8131
www.thepoultryfederation.com/

California Poultry Industry Federation
3117A McHenry Avenue
Modesto, CA 95350
888-822-4004
www.cpif.org

Delmarva Poultry Industry, Inc.
RD 6, Box 47
Georgetown, DE 19947
302-856-9037
www.dpichicken.org

Florida Poultry Federation
4508 Oak Fair Blvd., Suite 290
Tampa, FL 33610
813-628-4551
www.fl-ag.com

**International Foodservice
Distributors Association**
201 Park Washington Court
Falls Church, VA 22046
703-532-9400
http://www.ifdaonline.org

**Food Processing Machinery &
Supplies Association**
200 Daingerfield Road
Alexandria, VA 22314
800-331-8816
www.fpmsa.org

Food Safety Consortium
110 Agriculture Building
University of Arkansas
Fayetteville, AR 72701
501-575-5647
www.uark.edu/depts/fsc/index.htm

Georgia Poultry Federation
P.O. Box 763
Gainesville, GA 30503
770-532-0473

Indiana State Poultry Association
Purdue University
1151 Lilly Hall, Room G117
West Lafayette, IN 47906
765-494-8517
ag.ansc.purdue.edu/ispa

Kansas Poultry Association
Kansas State University
Dept. of Animal Sciences
139 Call Hall
Manhattan, KS 66505-1600
785-532-1201
www.oznet.ksu.edu

Kentucky Poultry Federation
P.O. Box 21829
Lexington, KY 40522
606-266-8375
www.kypoultry.org

Louisiana Poultry Federation
214 Knapp Hall
Louisiana State University
Baton Rouge, LA 70803
225-388-2219
www.lsuagcenter.com

Midwest Poultry Federation
108 Marty Drive
Buffalo, MN 55313-9338
763-682-2171
www.midwestpoultry.com

Mississippi Poultry Association
P.O. Box 13309
Jackson, MS 39236
601-355-0248
www.mspoultry.org

Missouri Poultry Federation
225 E. Capitol Avenue
Jefferson City, MO 65101
573-761-5610
www.thepoultryfederation.com

**National Association of Meat
Processors**
1910 Association Drive, Suite 400
Reston, VA 20191-1547
800-368-3043
www.namp.com

**National Association of State
Depts. of Agriculture**
1015 15th Street NW, Suite 930
Washington, D.C. 20005
202-296-2622
www.nasda.org

National Chicken Council
1015 15th Street NW, Suite 930
Washington, D.C. 20005-2622
202-296-2622
www.nationalchickencouncil.com

National Grocers Association
1825 Samuel Morse Drive
Reston, VA 20190
703-437-5300
www.nationalgrocers.org

**National Poultry & Food
Distributors Association**
3150 Highway 34 East
Suite 209, PMB 187
Newnan, GA 30265
770-535-9901
www.npfda.org

National Renderers Association
801 N. Fairfax Street, Suite 205
Alexandria, VA 22314
703-683-0155
www.nationalrenderers.org

National Restaurant Association
1200 17th Street NW
Washington, D.C. 20036
800-424-5156
www.restaurant.org

National Turkey Federation
1225 New York Avenue NW, Suite 400
Washington, D.C. 20005
202-898-0100
www.eatturkey.com

Nebraska Poultry Industry
A103 Animal Science
University of Nebraska Lincoln
P.O. Box 830940
Lincoln, NE 68583-0940
402-472-2051
nepoultry.org

North Carolina Poultry Federation
4020 Barrett Drive, Suite 102
Raleigh, NC 27609
919-783-8218
www.ncpoultry.org

Oklahoma Poultry Federation
P.O. Box 18938
Oklahoma City, OK 73154
(405) 229-5991
www.thepoultryfederation.com

Pacific Egg & Poultry Association
1521 I Street
Sacramento, CA 95814
916-441-0801
www.pacificegg.org

Packaging Machinery Manufacturers Institute
4350 N. Fairfax Drive, Suite 600
Arlington, VA 22203
703-243-8555
www.pmmi.org

Poultry Industry Council
483 Arkell Road,
Guelph, Ontario N1H 6H8
Canada
519-837-0284
www.poultryindustrycouncil.ca

Poultry Science Association
1111 N. Dunlap Avenue
Savoy, IL 61874
217-356-5285
www.poultryscience.org

South Carolina Poultry Federation
1921 Pickens Street, #A
Columbia, SC 29201
803-779-4700
www.scpoultry.org

South Dakota Poultry Industries Association
Animal and Range Sciences Department
South Dakota State University
Box 2170
Brookings, SD 57007
605-688-5409

Texas Poultry Federation
595 Round Rock West Drive, #305
Round Rock, TX 78681
512-248-0600
www.texaspoultry.org

U.S. Poultry & Egg Association
1530 Cooledge Road
Tucker, GA 30084
770-493-9401
www.poultryegg.org

United States Animal Health Association
P.O. Box 8805
St. Joseph, MO 64508
816-671-1144
www.usaha.org

Virginia Poultry Federation
333 Neff Ave # C
Harrisonburg, VA 22801
540-433-2451
www.vapoultry.com

ABOVE: Lyle, a Polish Crested pullet, takes a look at a visitor to the chicken coop at Brianne Riley and Matthew Taylor's home in Shelburne, Vermont. Riley became interested in chickens after working at Shelburne Farms as a Farmyard Educator when she was in college.

Appendix 3

State Resources

Alabama
Auburn University
Poultry Science Department
201 Poultry Science Building
260 Lem Morrison Drive
Auburn, AL 36849-5416
334-844-4133
www.ag.auburn.edu/ansc
www.aces.edu

Tuskegee University
Rm. 202 Morrison/Mayberry Hall
Tuskegee, AL 36088
334-727-1320
http://www.tuskegee.edu/Global/
story.asp?S=2406496

Alaska
Cooperative Extension Service
308 Tanana Loop, Room 101
University of Alaska Fairbanks
P.O. Box 756180
Fairbanks, AK 99775-6180
907-474-5211
www.uaf.edu/coop-ext

Arizona
University of Arizona
Forbes Building, Room 301
P.O. Box 210036
Tucson, AZ 85721
520-621-7205
www.ag.arizona.edu/extension

Arkansas
University of Arkansas
Poultry Science Department
Fayetteville, AR 72701
501-575-4952
www.poultryscience.uark.edu

Cooperative Extension Service
2301 South University Avenue
Little Rock, Arkansas 72204
501-671-2000
www.uaex.edu

**University of Arkansas
at Pine Bluff**
1890 Cooperative Extension Service
1200 N. University Drive
Pine Bluff, AR 71601
870-575-8000
www.uapb.edu

California
University of California at Davis
Animal Science Department
One Shields Avenue
Davis, CA 95616-8521
530-752-1250
http://animalscience.ucdavis.edu
http://animalscience.ucdavis.edu/
extension/index.htm

Colorado
Colorado State University
Campus Delivery 4040
Fort Collins, CO 80523-4040
970-491-6281
www.ext.colostate.edu

Connecticut
University of Connecticut
W.B. Young Building, Room 231
1376 Storrs Road, U-4134
Storrs, CT 06269
860-486-9228
www.cag.uconn.edu/ces/ces/index.html

Delaware
University of Delaware
Ulysses S. Washington Center
1200 N. DuPont Highway
Mail Code D160
Dover , DE 19901-2227
302-857-6424
ag.udel.edu/extension

Florida
University of Florida
Department of Dairy and Poultry
Sciences
P.O. Box 110920
Gainesville, FL 32610
352-392-1981
edis.ifas.ufl.edu/department_dairy_
and_poultry_sciences

Georgia
University of Georgia
Department of Poultry Science
217 Poultry Science Building
Athens, GA 30602
706-542-1325
www.poultry.uga.edu
www.caes.uga.edu

Hawaii
University of Hawaii at Manoa
1955 East-West Road
Agricultural Science 218
Honolulu, HI 96822
808-956-8234
www2.ctahr.hawaii.edu

Idaho
University of Idaho
Animal Veterinary Science
Department
Agricultural Science Building
Moscow, ID 83844-2330
208-885-6347
www.extension.uidaho.edu

Illinois
University of Illinois
214 Mumford Hall
1301 W. Gregory Drive
Urbana, IL 61801
217-333-5900
www.extension.uiuc.edu

Indiana
Purdue University
Department of Animal Science
915 W. State Street
West Lafayette, IN 47907-2054
765-494-4808
ag.ansc.purdue.edu/anscext
ag.ansc.purdue.edu/poultry

Iowa
Iowa State University
218 Beardshear Hall
Ames, IA 50011
515-294-4603
www.extension.iastate.edu/

Kansas
Kansas State University
123 Umberger Hall
Manhattan, KS 66506
785-532-5820
www.oznet.ksu.edu
www.asi.ksu.edu/DesktopDefault.
aspx?tabindex=651&tabid=229

Kentucky
University of Kentucky
College of Agriculture
S-107 Agricultural Science Building
- North
Lexington, KY 40546-0091
859-257-4302
www.uky.edu/Ag/AnimalSciences/
index.html
ces.ca.uky.edu/ces

Louisiana
Louisiana State University
Agricultural Center
P.O. Box 25203
Baton Rouge, LA 70894
225-578-4161
www.lsuagcenter.com/en/crops_
livestock/livestock/poultry/
www.lsuagcenter.com

Maine
University of Maine
5741 Libby Hall, Suite 102
Orono, ME 04469
207-581-2811
www.umext.maine.edu

Maryland
University of Maryland at College
Park
Department of Animal and Avian
Sciences
College Park, MD 20742
301-405-8746
ansc.umd.edu/extension/poultry/
index.cfm

Massachusetts
University of Massachusetts
Extension Office
Draper Hall
University of Massachusetts
Amherst, MA 01003
413-545-4800
www.umassextension.org

Michigan
Michigan State University Extension
Agriculture Hall, Room 108
East Lansing, MI 48824
517-432-3849
www.msue.msu.edu/portal

Minnesota
University of Minnesota
Coffey Hall, Room 240
1420 Eckles Avenue
St. Paul, MN 55108
612-624-2703
www.extension.umn.edu/index.html

Mississippi
Alcorn State University
1000 ASU Drive, #690
Lorman, MS 39096
601-877-6137
www.alcorn.edu
www.asuextension.com/asuep/index.php

Mississippi State University
Department of Poultry Science
P.O. Box 5188
Mississippi State, MS 39762
662-325-3416
www.msstate.edu/dept/poultry/
msucares.com/poultry/consumer/
index.html

Missouri
University of Missouri
309 University Hall
Columbia, MO 65211
573-882-7754
http://extension.missouri.edu

Montana
Montana State University
P.O. Box 172040
115 Culbertson Hall
Bozeman, MT 59717-2040
406-994-6647
www.montana.edu
www.animalrangeextension.
montana.edu

Nebraska
University of Nebraska
211 Agriculture Hall
Lincoln, NE 68583
402-472-2966
www.extension.unl.edu/home

Nevada
University of Nevada at Reno
Cooperative Extension
National Judicial College 118,
Mail Stop 404
Reno, NV 89557-0106
775-784-7070
www.unce.unr.edu

New Hampshire
University of New Hampshire
Cooperative Extension
Taylor Hall
59 College Road
Durham, NH 03824
603-862-1520
extension.unh.edu

New Jersey
Rutgers, The University of New Jersey
Rutgers Cooperative Extension
Martin Hall, Room 305
88 Lippman Drive
New Brunswick, NJ 08901
732-932-5000
njaes.rutgers.edu

New Mexico
New Mexico State University
Department 3AE
P.O. Box 30003
Las Cruces, NM 88003
505-646-3016
extension.nmsu.edu

New York
Cornell University
365 Roberts Hall
Ithaca, NY 14853
607-255-2116
www.cce.cornell.edu

North Carolina
North Carolina A&T State University
P.O. Box 21928
Greensboro, NC 27420
336-334-7691
www.ag.ncat.edu/extension

North Carolina State University
Extension Poultry Science
229 Scott Hall
Campus Box 7608
Raleigh, NC 27695-7608
919-515-2621
www.ces.ncsu.edu

North Dakota
North Dakota State University
315 Morrill Hall
P.O. Box 6050
Fargo, ND 58105
701-231-8944
www.ag.ndsu.edu/extension/

Ohio
Ohio State University Extension
Ohio State University
3 Agricultural Administration
Building
2120 Fyffe Road, Room 4
Columbus, OH 43210
614-292-6181
ohionline.ag.ohio-state.edu

Oklahoma
Oklahoma State University
Division of Agricultural Sciences and
Natural Resources
136 Agricultural Building
Stillwater, OK 74078-6051
405-744-5398
www.ansi.okstate.edu/exten
www.dasnr.okstate.edu

Oregon
Oregon State University
Department of Animal Science
101 Ballard Extension Hall
Corvallis, OR 97331-6702
541-737-5066
oregonstate.edu/outreach/
extension.oregonstate.edu

Pennsylvania
Pennsylvania State University
Department of Poultry Science
213 William N. Henning Building
University Park, PA 16802
814-865-3411
poultry.cas.psu.edu
poultryextension.psu.edu

Rhode Island
University of Rhode Island
9 E. Alumni Avenue, Room 137
Kingston, RI 02881
401-874-2970
www.uri.edu/ce/index1.html

South Carolina
Clemson University
Animal and Veterinary Sciences
Department
129 Poole Agricultural Center
Box 340361
Clemson, SC 29634-0361
864-656-3427
cufp.clemson.edu/avs/
virtual.clemson.edu/groups/public

South Dakota
South Dakota State University
Agricultural Hall 154
P.O. Box 2207D
Brookings, SD 57007
605-688-4792
sdces.sdstate.edu/

Tennessee
The University of Tennessee
121 Morgan Hall
P.O. Box 1071
Knoxville, TN 37901
865-974-7114
www.utextension.utk.edu

Texas
Texas A&M University
Poultry Science Department
Kleberg Center, Room 101
College Station, TX 77843-2472
979-845-1931
gallus.tamus.edu/departmental.html

Utah
Utah State University
4900 Old Main Hill Road
Logan, UT 84322
435-797-2200
extension.usu.edu

Vermont
University of Vermont
College of Agriculture and Life
Sciences
601 Main Street
Burlington, VT 05405
802-656-2990
ctr.uvm.edu/ext

Virginia
Virginia Polytechnic Institute and
State University
Department of Animal and Poultry
Sciences
101 Hutcheson Hall
Blacksburg, VA 24061-0306
540-231-5299
www.apsc.vt.edu
www.ext.vt.edu

Washington
Washington State University
421 Hubert Hall
P.O. Box 646230
Pullman, WA 99164
509-335-4561
ext.wsu.edu

West Virginia
West Virginia University
P.O. Box 6031
Morgantown, WV 26506-6031
304-293-5691
www.wvu.edu\~exten

Wisconsin
University of Wisconsin
Department of Animal Sciences
Animal Science Building
1675 Observatory Drive
Madison, WI 83706-1284
608-263-4300
www.uwex.edu/ces

Wyoming
University of Wyoming
P.O. Box 3354
Laramie, WY 82071
307-766-5124
ces.uwyo.edu

ABOVE: An Australorp rooster waits in his cage to be taken to his new home during the Vermont Bird Fanciers Club quarterly swap held in East Randolph, Vermont.

Appendix 4

Poultry Hatcheries and Supplies

American Livestock Breeds Conservancy
Rare Breeds, Breeders, and Products Directory
P.O. Box 477
Pittsboro, NC
27312 919-542-5704
albc@albc-usa.org; www.albc-usa.org

A nonprofit organization dedicated to the conservation and promotion of breeds of livestock and poultry threatened with extinction. For those people serious about breed conservation, this is a great source of the top breeders of heritage poultry in the United States.

Eldon's Jerky & Sausage Supply
P.O. Box 422 022
Main Street
Kooskia, ID 83539
800-352-9453
FAX 208-926-4949
www.eldonsausage.com

Okay, so they don't sell chicks or eggs, but they do have every kind of supply you can imagine necessary to make scrumptious chicken sausage.

Hall Brothers Hatchery
P.O. Box 1026
Norwich, CT 06360
860-886-2421

A small family hatchery where I bought my very first meat chickens. They tend to have hybrid varieties rather than the more obscure breeds.

Moyer's Chicks, Inc.
266 East Paletown Road
Quakertown, PA 18951
215-536-3155
www.moyerschicks.com

Another great source for hybrid meat and laying chicks.

Murray McMurray Hatchery
Box 458
191 Closz Drive
Webster City, IA 50595-0458
800-456-3280
www.mcmurrayhatchery.com

Touted as America's rare-breed hatchery, this is arguably the most well known hatchery in the United States It is also one of the more expensive hatcheries and really just a broker for a number of different breeders. The color catalog, complete with real pictures of the chicks for each breed, is a good place to begin, especially if you're not sure which breed you want to start with. Their service and quality are also impeccable.

Nasco Farm & Ranch
901 Janesville Avenue
Fort Atkinson, WI 53538
800-558 – 9595
www.enasco.com

This supplier has everything from feed to sausage-making supplies.

Premier Fence Supply
P.O. Box 89
Washington, IA 52353
800-282-6631
www.premier1supplies.com

They have electroplastic netting for poultry yards and fence chargers.

Reich Poultry Farms, Inc.
1625 River Road
Marietta, PA 17547
866-365-0367 or 717 426-3411;
FAX 717-426-8061

This is one of my favorite hatcheries for hybrid meat and egg varieties. They supply large poultry barns with thousands of chicks at a time, but they know me by name when I call up to order my 150 Buff Silver meat birds. I have had great success with several of their hybrids on forage.

Sand Hill Preservation Center
1878 230th Street
Calamus, IA
52729-9659
www.sandhillpreservation.com

A small, family-run farm and company dedicated to the preservation of genetic resources. Their poultry offerings are outstanding, and although you'll probably have to exercise delayed gratification before you get your birds (they do everything by mail, and there are usually long waiting lists for the rarer breeds), you won't be disappointed.

Townline Hatchery
P.O. Box 108
Zeeland, MI 49464
616-772-6514
www.townlinehatchery.com

Fast-growing, traditional hybrid meat birds, hybrid laying hens, and turkeys. Very good stock for serious meat-producing potential.

ABOVE: Noah Abbot, 8, of Randolph, Vermont, Silas Mitchell, 8, of Orwell, Vermont, and David Russell of Orwell, Vermont, play with an Old English Bantam rooster, left, and hen brought by the author, right, to a workshop she gave to children of parents attending the NOFA-VT Annual Conference in Randolph Center, Vermont.

ABOVE Clucky, a Partridge Cochin pullet, roams amongst the leaves at Brianne Riley and Matthew Taylor's home. The couple raised their menagerie of a dozen chickens from chicks they bought from a hatchery.

Appendix 5

Recommended Reading

Some of the publications listed below are constantly evolving and getting updated. One of the easiest ways to find the latest information is to do a Web search using the previous edition's title or organization name.

ABC of Poultry Raising (1977), by J. H. Florea, Dover Publications Inc., 31 East 2nd Street, Mineola, NY 11501. This book contains many useful plans for poultry feeders, nest boxes, housing, yards, etc., including lists of materials needed for their construction.

American Standard of Perfection (latest edition), American Poultry Association, c/o secretary-treasurer, Lorna Rhodes, 133 Millville Street, Mendon, MA 01756; 508-473-7943; www.amerpoultryassn.com. This is a necessary guide for those wishing to exhibit and breed poultry, with complete descriptions of all recognized breeds and varieties of domestic poultry.

Animal, Vegetable, Miracle (2007), by Barbara Kingsolver, HarperCollins Publishers, 10 East 53rd Street, New York, NY 10022. While I had previously devoured all of Kingsolver's non-fiction works and loved every one, I enjoyed her description of her family's yearlong adventure and their attempts at self-sustenance even more. This is an easy read, it's practical, and there are some great recipes by her daughter Camille and thoughtful essays by her husband Steven L. Hopp mixed in along the way.

Basic Butchering of Livestock & Game (1986), by John Mettler, Jr., DVM, Storey Communications, Inc., Pownal, VT 05261. My copy of this book is threadbare from use. It has butchering techniques for everything from moose to turkey. The illustrations are incredibly helpful if you're new at this.

Charcuterie (2005), by Michael Ruhlman and Brian Polcyn, W.W. Norton & Company, Inc., 500 Fifth Avenue, New York, NY 10110. This is an all-around great book for salting, smoking, and curing all kinds of meats.

Chicken Tractor (1994), by Andy Lee, Good Earth Publications, P.O. Box 898, Shelburne, VT, 05482. A back-to-the-land approach to raising chickens on pasture to improve your soil and farmstead.

FeatherSite, www.feathersite.com. Full descriptions of hundreds of chicken breeds, often including actual photographs and club and private breeder contacts.

Fields of Plenty (2005), by Michael Ableman, Chronicle Books LLC, 85 Second Street, San Francisco, CA 94105. A wonderful description of Ableman's journey across the United States as he looks at several different aspects of American agriculture. A great book for people who enjoy growing and eating food.

How to Cook Everything (1998), by Mark Bittman, MacMillan Publishing USA, 1633 Broadway, New York, NY 10019. Everyone in my family, including my nine-year-old son and the seasonal apprentices who come to the farm to learn how to grow and cook food, sometimes for the very first time, reach for this book. It's easy to read and easy to follow.

The Chicken Health Handbook (1994), by Gail Damerow, Storey Communications Inc., Schoolhouse Road, Pownal, VT 05261. An in-depth guide to preventing, identifying, and treating diseases common to chickens.

The Fairest Fowl (2001), by Tamara Staples and Ira Glass, Raincoast Books, 9050 Shaughnessy Street, Vancouver, BC, V6P 6E5, Canada. A fun, coffee-table book, with portraits of championship chickens and accompanying natural history facts.

The Field Guide to Chickens (2006), by Pam Percy, Voyageur Press, MBI Publishing Company, Galtier Plaza, Suite 200, 380 Jackson Street, St. Paul, MN 55101-3885. This field guide details more than 60 breeds, with anecdotal history and physical characteristics of the breeds.

The Omnivore's Dilemma (2006), by Michael Pollan, Penguin Group (USA) Inc., 375 Hudson Street, New York, NY 10014. Read this book. It will change the way you think about food.

The Unsettling of America (1977), by Wendell Berry, Sierra Club Books, 530 Bush Street, San Francisco, CA 94108. Berry offers a review of agriculture and cultural practices which have lead to the loss of living with the land and steps to return to good farming practices. If you're new to his writing, this is a great book to start with.

Poultry Press, P.O. Box 542, Connersville, IN 47331; 765-827-0932; www.poultrypress.com. A great periodical for chicken fanatics.

Poultry Production (1935), by William Lippincott and Leslie Card, Lea & Febiger, Philadelphia, PA 19019. This is considered by many to be the definitive source for poultry production. A fascinating read for poultry enthusiasts, comparing modern methods with past production.

Small Farmer's Journal, Lynn R. Miller, Editor, Small Farmer's Journal, Inc., 192 W. Barclay Drive, Sisters, Oregon 97759. Subscription information at SFJ, P.O. Box 1627, Sisters, Oregon 97759. This journal's mantra is "There's never been a better time to be a horsefarmer," though the magazine has practical small-farm information for anyone interested in diverse, small-scale agriculture. Published quarterly, I read it cover to cover when it arrives in the mail.

ABOVE: The author at age 6 (left) and her sister, Anna, holding their Rhode Island Red pullets, Penelope and Hortense.

Acknowledgments

There are many humans that I would like to thank for helping me write of personal experiences with my favorite barnyard creatures, chickens. My parents understood that it was far easier for me to relate to the antics of the chickens as a troubled teenager than to attempt to fit in with the beautiful people at school, whom I envied but did not understand. My husband has indulged me with countless pets who have done nothing more than cheer me at their sight, not even required to contribute an egg or a drumstick to the farm to ensure their continued existence. My sisters both encouraged me to pursue writing, and Anna, especially, provided me with more than just ideas for some of my anecdotes.

I thank those who patiently pored over these pages, especially Ann Treistman, an editor who allowed me to complete this project despite being occupied with lambing, haying, and the growing season, and Jane Crosen, whose insightful comments helped me immensely. Geoff Hansen, thank you for suggesting this project, and thanks, too, for allowing me to be associated with your incredible art.

And to the chickens: to Taurus, Funny Face, Fern, Bright Eyes, Pica Pica, Wendell Berry, Forest Gump, Peggy, Danny, Henry, Juan Carlos Domingues, Poopie Poo, and House Hen. Thank you for reminding me that we're all part of the same book, and that I'm just one small part in the chapter.

—Jennifer Megyesi

I'm very thankful to those who allowed me to spend time photographing the various ways to keep chickens: Karl Hanson and Cloë Milek, Jennifer Hauck and Alex Cherington, Suzanne Long and Tim Sanford, Ralph and Cindy Persons, Brianne Riley and Matthew Taylor, Rick Schluntz and Carol Steingress, Ray Williams and Liz York, and the Northeast Poultry Congress' organizers and participants. I'm especially grateful for working with Jennifer Megyesi, who was always helpful with suggestions and was a great collaborator on this book. Thanks to the crew at Skyhorse Publishing, including Ann Treistman, Abigail Gehring and LeAnna Weller Smith, who put all of the pieces together into one special book.

—Geoff Hansen

Index